# Studies in Computational Intelligence

## Volume 504

*Series Editor*

Janusz Kacprzyk, Warsaw, Poland

For further volumes:
http://www.springer.com/series/7092

Mirsad Hadžikadić · Sean O'Brien
Moutaz Khouja
Editors

# Managing Complexity: Practical Considerations in the Development and Application of ABMs to Contemporary Policy Challenges

 Springer

*Editors*
Mirsad Hadžikadić
College of Computing and Informatics
Software and Information Systems
University of North Carolina
Charlotte
North Carolina
USA

Moutaz Khouja
The University of North Carolina
Charlotte
North Carolina
USA

Sean O'Brien
Strategic Analysis Enterprises
Fairfax Station
Virginia
USA

ISSN 1860-949X          ISSN 1860-9503 (electronic)
ISBN 978-3-642-43598-0  ISBN 978-3-642-39295-5 (eBook)
DOI 10.1007/978-3-642-39295-5
Springer Heidelberg New York Dordrecht London

# Contents

# Chapter 1
# Addressing Complex Challenges in an Interconnected World

Sean O'Brien[1], Mirsad Hadžikadić[2], and Moutaz Khouja[2]

[1] Strategic Analysis, Enterprises, Fairfax, Station, VA, USA
[2] University of North Carolina, Charlotte, NC, USA

## 1.1 The World around Us: Constant Change, Uncertainty, Chance, and Evolution

As we were growing up, the world seemed very orderly. Adults knew what was right and what was wrong (and they made sure they told as that). Television "talking heads" explained matter-of-factly the causes and consequences of the events of the day in no uncertain terms. Politicians peddled their ideas for solving all problems of interest to voters. Scientists were convinced that they discovered the smallest elements of matter, i.e. atoms. Economists staunchly defended the notion that markets were efficient and that people were completely rational. There were no environmentalists to speak of. The weather the next day, of course, was going to be just a slight variation of the current weather conditions.

Then, we grew up and suddenly realized that the world is a constantly evolving arena where shifting political alliances, cyclical economies, new scientific discoveries, frequent natural and man-made catastrophes, fluid relationships, changing opinions, and self-motivated perceptions of reality are the norm rather than the exception. In such a world, in order to succeed, individuals and organizations (associations, corporations, and governments) must constantly observe, assess, and adjust to the environment that is, at the same time, changed by the very action of each of its constituent agents.

Consequently, the list of challenges with potentially catastrophic consequences for our way of life, our nation, our economy, and even civilization in general, seem to accumulate much faster than the ability of our public leaders to develop a consensus behind their causes, never mind their resolution. Scarcely a day passes in which we are not bombarded with reminders that we live in dangerous times. Global warming, cyber attacks on critical infrastructure, threats to our food and water supply, global health pandemics, drug and human trafficking, nuclear weapons proliferation, global financial crises, and unspeakable atrocities carried out by terrorists and government agents against our fellow human beings are just a few of the daily headlines that occupy our 24 hour global news cycle.

M. Hadžikadić (eds.), *Managing Complexity*,
Studies in Computational Intelligence 504,
DOI: 10.1007/978-3-642-39295-5_1, © Springer-Verlag Berlin Heidelberg 2013

In *Turbulence in World Politics: a Theory of Change and Continuity*, a Choice Award Winning book of 1991, James N. Rosenau takes us on a journey to explain how we got to this point. According to Rosenau, until the 1950s citizens were generally uneducated and spent much of their time toiling in labor and raising their families. Their governments controlled or manipulated what little information was available to them. People did not challenge or question government authority. They complied out of habit. Most of the problems or challenges a country faced occurred within the country itself, and governments generally dealt with these challenges effectively. Governments could conduct international affairs largely as they wished. To the extent they required public support for their actions, they could manufacture that support through skillful manipulation of the media, which operated on a much slower news cycle than today's news organizations. In those days, civil society was generally weak, and the state system reigned supreme.

After 1950, this bifurcated system began to fragment. The global financial and economic systems became increasingly interdependent. Because large corporations could locate their headquarters and operations almost anywhere (and nowhere in particular in the case of internet-based services), governments could no longer exercise exclusive control over large corporations, making it much more challenging to tax or regulate these organizations. Air travel increased both the speed with which people could cross borders to transact legitimate business, engage in tourism, traffic in people or drugs, or deliver weapons against their enemies. If any development ended the notion that problems could be contained within countries it was the birth of nuclear weapons that threatened the lives of individuals in every corner of the world.

Other problems, such as global warming defied consensus explanations, easy solutions, or the ability of any single country or even groups of countries to address the challenge in any meaningful manner. Some of the more advanced industrialized countries--those that had both generated the most harmful pollutants and reaped the most economic gains—frequently demanded that other less developed, but newly industrializing countries, first restrain their generation of harmful byproducts and, therefore, their economies, before doing so themselves. These demands raised fundamental questions about sovereignty, fairness, relative contributions, self-determination, and national self-restraint.

According to Rosenau, one of the most decisive developments in the post-1950s eras that altered the balance between strong governments and weak societies was the microelectronics revolution that eventually gave birth to the Internet. As media organizations proliferated and the Internet matured, the ability of governments to control and manipulate images and information to their advantage was severely eroded. As governments struggled to deal with problems—many originating beyond their borders—the media were all too eager and prepared to bring images to televisions, print media, and computer screens to viewers around the world, who were becoming increasingly sophisticated in their skills to process and act upon those images. As media organizations proliferated, so too did a plethora of special interest organizations that pressured or threatened their governments to address problems and interests that were increasingly beyond their control. These cross-pressures from diverse interests and constituencies made it even more difficult for governments to respond to challenges effectively, especially those leaders subject to regular democratic elections.

The 21$^{st}$ century has exacerbated the problems even further. Advances in communications, information, transportation, energy, and financial technologies have integrated the world to such a degree that it is virtually impossible today for any country to be fully protected with its own means in the context of economic, political, social, or military upheavals. The connections that make goods, information, and people flow around the world have finally created a web that makes us all part of one system. Today, ideas, threats, viruses, and missiles can arrive in the blink of an eye. The distribution of their arrivals often follows the Power Law, which defines systems of non-linear, self-organizing, and emergent nature. Such systems are usually referred to as complex systems.

## 1.2    Complex Adaptive Systems: The Case of the European Debt Crisis

Many of the challenges we face in society occur within the context of a world that behaves as a Complex Adaptive System (CAS). A CAS consists of many interconnected parts, e.g., people, countries, organizations, or computer systems. The parts are adaptive in that they have the ability to learn from their experiences and results of activities, i.e. their successes and failures.

The European Union (EU) is a prime example of a CAS. The EU is currently in the 3rd year of a debt crisis, whose contagion to other countries now poses one of the largest threats to the global economy since the Great Depression. A brief examination of this crisis is instructive for illustrating the challenges in studying complex adaptive systems. The EU is comprised of 17 independent countries that share a common currency, the Euro. Although the original vision behind the EU was that it would one day evolve into a United States of Europe, this vision has not been achieved and the common currency is under constant pressure. Each EU country remains independent and responsible for its own debt and spending activities. Many of them still must submit proposed changes to their status in the common currency to their elected parliaments for approval. In some countries (e.g., Ireland), changes to their status within the EU must be submitted to a popular referendum. Although countries in the Eurozone have pledged to maintain their debt-to-GDP ratios at a pre-determined level, some countries have not been able to do so to this day.

The "Eurozone Crisis" began in 2009, when it was revealed that the Greek government had lied about its debt level, which had ballooned to a projected 198% of its GDP. As investors began to doubt the ability of Greece to repay its debts, given that burden, investors in government debt began to demand a higher premium to lend to the government. Interest rates on those bonds increased to levels that further increased pressure on the ability of the Greek government to repay its debt. But investor fears did not stop with Greece. They realized that if Greece had a high chance of defaulting on its debts, other countries facing similar structural challenges might also struggle to repay investors. So investors began to demand higher premiums on bonds issued by Ireland, Portugal, Spain, and Italy, thus further stressing the financial systems in those countries as well.

The higher interest rates pushed these countries into even more precarious economic conditions. Greece was the first country to request an international bailout, followed by Ireland and Portugal. Fiscal authorities, backed by Germany, the strongest economic member of the EU, demanded that in exchange for bailing these countries out, they would have to regain control of their public finances by implementing stringent austerity measures that included a combination of drastic reductions in public spending and increased taxes. Germany, backed by European economic authorities, argued that reining in profligate spending was the only responsible way to restore investor confidence and place these countries back on a path of economic sustainability.

But as some governments slashed public spending their unemployment rates increased, which in turn made it increasingly difficult for people to purchase basic goods and housing, reduced spending, and resulted in economic contraction. When the housing bubble burst in Spain, many of the region's largest banks were left holding bad mortgage securities that threatened their viability. As unemployment increased throughout the region—going as high as 25% in some countries—spending on goods ground to a halt, and people took to the streets to protest against the German-backed austerity measures. Extreme right and left political parties, who opposed the bailout terms, gained strength at the expense of more centrist parties which included technocrats with experience in managing economic turbulence.

Because of the interconnectedness of the global economy, these economic troubles could not be contained within the Eurozone itself. About 20% of all US exports go to countries in the 17-member Eurozone, and a much higher proportion of Chinese goods are destined for the region. As trade declined and uncertainty increased, US companies restrained their hiring, thus threatening to slow the US recovery from its own recession. China's exports to the region abruptly halted, imperiling the government's plans to engineer a soft landing for its own overheated economy.

The Eurozone crisis, and its threat to the global economy, has stimulated an intense debate in capitals around the world as to its causes and potential resolution. Many economists and politicians argue that austerity measures—drastic reductions in government spending and increases in taxes—are the wrong recipe for countries struggling on the precipice of recessions. Lower spending and higher taxes take liquidity out of the economic system and retard productive investments that could generate economic growth. What is needed, these economists argue, is precisely the opposite—short term *increases* in government spending and *lower* taxes to spur economic growth. As the affected economies improve, governments will generate higher revenue that can then be dedicated to reducing their long-term debts. Those who subscribe to this view argue that austerity measures have made matters worse in the Eurozone.

Other economists and policymakers draw the opposite conclusion. They argue that the Eurozone countries are failing because they spent too much and taxed too little. Those who favor drastic reductions in spending in the US point to Greece as an example of what will happen if the US fails to reign-in its budget deficit. Here we have two very different perspectives on what caused this global economic crisis, and two very different prescriptions for how it might be resolved. The fate of the global economy could very well hinge on our understanding of which view is the correct one.

## 1.3   Agent-Based Models

The Eurozone crisis, like so many policy challenges that affect the world today, poses significant challenges for analysis. How might we evaluate the consequences of alternative sequences of complicated policy responses, when so many firms, institutions, countries, and special interest groups are involved? Comparative case study approaches are ill equipped to deal with so many interconnected parts and variables. Econometric analyses have extensive data requirements, have difficulty analyzing dynamic phenomena across multiple levels of analysis, and do not allow for analyzing potential states that are not explicitly represented in the system.

Agent-based models, by contrast, have emerged in recent years as an alternative method for evaluating the implications of both strategic and tactical interactions among individuals, organizations, and countries. Agents are virtual entities that are encoded with features designed to represent the key actors in a system that is to be modeled. Agents can represent individuals, groups of individuals, firms and institutions, political leaders, or any other entity that can be ascribed a set of characteristics and motivations (in this case, in social environments). Agents are purposeful, goal-oriented, and situated in networks much like the real world entities that they represent. They are programmed to interact with other agents and, based on the response, take actions designed to achieve their goals. From simple rules that govern micro-level interactions a wide variety of higher order, macro-economic/social/cultural phenomena emerge.

Agent-based models can be used to assess how different states of a system might evolve, or how changing agent motivations or characteristics can alter those states. As such, if designed in a disciplined manner, agent-based models can be used to evaluate how alternative policy responses to changes in the parts of a system can affect the system as a whole. There are several notable examples of how agent-based models can be constructed to evaluate the effects of various policy approaches to complex problems:

1.  In 1996, Sandia Labs developed ASPEN, an agent-based model of the US economy. Researchers created agents that represented households, firms, and

governmental institutions like the Federal Reserve Bank. The rules that motivated agents and governed their interactions were based on both macro and micro-economic theories, and were designed to be consistent with the way in which the US economy operates. To validate the model, researchers examined how actual policy changes, such as inflationary or deflationary monetary policies in the 1970s, affect agent behaviors and the economy as a whole. Developers then compared the model-projected results with historical accounts of the effects of these policies to assess the adequacy of the model for generating useful insights.

2. Lustick and Eidelson (2004) were among the first political scientists to publish an analysis, in the *American Political Science Review,* that used an agent-based model to evaluate the veracity of theories of ethnic secessionism that had been subjected to competing claims. Lustick and his colleagues noted that neither large nor small sample studies had been able to provide a consensus on whether power sharing prevented or encouraged secessionism in multi-cultural societies. The authors developed an agent-based model (called Beita) that was designed to be theoretically consistent with any multi-cultural state. By controlling some aspects of the experimental environment, while varying others, the researchers were able to perturb the system in ways consistent with many of the competing theories. This allowed them to determine the conditions under which each contending theory was valid.

3. Scholars at Virginia Tech developed the first 100-million agents simulation that, based on census and other data, embodied synthetic, empirically representative versions of most American citizens. Each agent had upwards of 163 different encoded features. Sponsored by several US government agencies, the system was used to evaluate how disease pandemics could spread throughout the country, and what effects different policy responses would have on minimizing the damage (O'Brien, 2010).

## 1.4   About This Book

This book emerged out of a project initiated and funded by the Defense Advanced Research Projects Agency (DARPA) that sought to build on efforts to transform agent-based models into platforms for predicting and evaluating policy responses to real world challenges around the world. It began with the observation that the social sciences literature is replete with theories of human behavior whose validity has yet to be tested. These theories remain subject to conflicting assertions. Theories of human behavior are often used to estimate the consequences of alternative policy responses to important issues and challenges. However, alternative theories that remain subject to contradictory claims are ill suited to inform policy.

The vision behind the DARPA project was to mine the social sciences literature for alternative theories of human behavior, and then formalize, instantiate, and integrate them within the context of an agent-based modeling system. Software agents were to be created and empirically endowed with the characteristics of the real world people and institutions they were designed to represent. By creating a large, historical dataset of events and conditions around the world, we planned to use the experimental platform—much like Lustick and Eidelson (2004) used their Beita system—to evaluate the conditions under which alternative theories and groups of theories applied. The end result was to be a proof of concept—developed from the ground up—of social knowledge that could be used as a better guide for policy analysis than what was generally acknowledged to be the state of the art at the time.

This book describes in detail the approach we took to defining and implementing a pilot system that helped DARPA assess the feasibility of a computational social science project on a large scale. Chapter 2 provides an overview of the concept of complexity that has been the driving perspective of our understanding of the issue of social change in the context of political and military upheaval. Chapter 3 offers a description of agent-based methodology, the most widely used method for implementing complexity principles. Chapter 4 builds on Chapter 3 by focusing on the use of agent-based models to explain past events, anticipate possible futures, and evaluate intervening policies for the system under consideration. With this theoretical background in mind we then turn, in Chapters 5 and 6, to the practical example of the "Actionable Capability for Social and Economics Systems" (ACSES), a DARPA-funded prototype of the battle for the hearts and minds of the population in the specific region of Afghanistan in the post 9/11 world. These chapters also address the issue of defining, implementing, and evaluating specific policies addressing specific challenges. Chapter 7 then takes this specific example and generalizes it in a sequence of steps that needs to be followed in order to build successful agent-based modeling applications. Chapter 8 concludes with a list of lessons learned both in the ACSES project and the subsequent analysis of successes and failures associated with this project. Finally, Chapter 9 provides some ideas for advancing social science through agent-based modeling.

# Chapter 2
# Complexity: Where Does It Come from?

Sean O'Brien[1], Mirsad Hadžikadić[2], and Moutaz Khouja[2]

[1] Strategic Analysis, Enterprises, Fairfax, Station, VA, USA
[2] University of North Carolina, Charlotte, NC, USA

## 2.1 Viewing the World as a Complex System

Chapter 1 describes our world as fluid, dynamic, and uncertain. The prudent question to ask then is: What makes it so? In order to answer this question we first need to look at the architecture of the world itself, regardless of whether we are talking about the natural world or the social one.

All systems consist of constituent elements, although our understanding of these elements changes over time. For example, scientists still do not know the nature of the most elementary particles in the physical world. Early on scientists posited that the world consisted of four elements: fire, water, air, and earth. Later on they discovered atoms, only to realize that atoms consist of even smaller particles, namely electrons, protons, and neutrons. Modern science now suggests that there exist sub-atomic particles, including quarks, leptons, and gauge bosons. Regardless of this lack of our complete understanding of the basic building blocks of the world we live in, it is still wildly believed that such building blocks exist in some form.

However, do we really need to understand quarks before we can proceed with learning how to drive a car? Of course not—learning to drive involves our understanding of the functions of the steering wheel, the gas pedal, or the breaks, not the quarks. The point is, when we talk about elements of a specific system we think of things, phenomena, behavior, or events that are at the level that is sufficient for our effective reasoning about the system itself.

Once we decide on the most relevant elements of the system, whatever they are, we immediately realize that these elements interact with each other. Rocks rest on other rocks, genes influence the production of proteins, and people talk to each other in person, via Facebook, or using some other communication-enabling gadget. Obviously, the nature of these interactions gets increasingly complicated as we move from the physical to the social world. This is reflected in our ability to quantify the physical world much more readily than we are capable of quantifying the social world, where the existence of free will makes things much less predictable.

M. Hadžikadić (eds.), *Managing Complexity*,
Studies in Computational Intelligence 504,
DOI: 10.1007/978-3-642-39295-5_2, © Springer-Verlag Berlin Heidelberg 2013

In the end, it is these interactions that make our world (and life within it) so interesting, uncertain, and unpredictable. One of the characteristics of complex systems is that they cannot be understood just by reducing them to their constituent elements. Doing so would cause the loss of understanding of how these elements interact with each other and the implications of these interactions. This is important because it is these interactions that are responsible for producing the aggregate, emergent behavior that we observe at the system level. The saying "The whole is greater than the sum of its parts" best captures this phenomenon. It is also interesting to note that this relationship between systems and their components is bidirectional in that it is this behavior of the components that results in aggregate, emergent properties at the system level, which in turn affects the behavior of the components themselves.

The component interactions are "two-way" streets as well, as components influence each other simultaneously. Consequently, each component receives feedback either directly from their peers or indirectly from the constraints and norms established by the system. This feedback can be either negative or positive. Both types of feedback are important, as negative feedback dampens undesirable behavior while positive feedback encourages repetition of desirable actions.

By influencing each other components actually co-evolve in time. Evolutionary biologists have demonstrated that prey species become more efficient as they attempt to match improvements in their predators. Similar changes happen with people, organizations, and countries as they use education, science, technology, innovation, and political alliances to position themselves favorably in this constant struggle for the survival of the fittest. The shifting fitness landscapes at the system level, caused by the constantly co-evolving components, make this struggle even more demanding and uncertain.

These co-evolving components and shifting fitness landscapes combine to produce systems where threshold effects abound. The existence of such threshold effects indicates that complex systems are non-linear in nature. This means that some phenomena just simmer under the surface, until a possibly unrelated event helps it cross the magic barrier of resistance and develop into an overwhelming force. Wars, cancer, earthquakes, tornados, rush-hour traffic, stock market crashes, and epidemics are but a few examples of such phenomena. Threshold phenomena have been studied in many disciplines, most notably in physics under the name of phase transition.

The saying *"One never steps in the same river twice"* summarizes the above discussion rather well. Even though the world feels familiar to us on a daily basis, it is never truly the same for two reasons: (1) the constant actions of all components, and (2) our own reaction to the actions of others. This realization that the world is inherently dynamic helps each component appropriately adjust to the world it exists in.

## 2.2   How Should We Deal with a Complex World?

There are two approaches to modeling the world around us in order to better understand it. One approach is known as the *reductionist* or top-down approach. In the reductionist approach the modeler assumes that all problems can be reduced to their constituent components, without any loss of information relevant to the problem-solving process. Then, the modeler gains knowledge of the nature of each component by creating a model of each component independent of the models of other components. Sometimes, the modeler needs to go deeper into the structural hierarchy of components, as each of them can be a system of components in its own right.

Once the components comprising the system are well understood, the modeler puts the system back together and then analyzes the overall system. This process works well for problems/systems whose components interact either loosely/ infrequently or predictably. For example, components that make up a car or an airplane have predictable and well-understood ways in which they interact with each other. Hence, learning how to design each car component and how such components affect each other will lead to a system that performs according to its technical specifications. When a car fails, we know with a fairly high certainty which parts could have caused the failure. Since the behavior of such elaborate systems is well understood and predictable, we call them *complicated* instead of *complex* systems. They are characterized by known distributions, average values, and stable rules of the game.

However, many systems involve component interactions that are often unpredictable, flexible, dynamically changing, and contextual. We have already mentioned many of such examples, including economy, brain, cancer, war, or ecology. These phenomena are examples of what we mean by complex systems: systems whose nature is defined by the nature of the ever-changing interactions among the system's constituent parts. These not-so-well-understood, complexity-driven phenomena often must be modeled from the bottom up in order to capture the essence of their nature. This means that the modeler must first model the system components. Next, the modeler must define all possible actions for each component, whether they refer to the behavior or the physical movement of components. How each component behaves depends on both the "genetic" material of the component and its history of interactions with other components. Because of the clear analogy with social systems, such components are often called "agents," hence the term "agent-based modeling."

Once the agents and their behavior are established within the system environment, the modeler runs numerous simulations of the model to see what kind of a system *emerges* on average from this collection of agents and their possible behaviors. Things get quickly complicated as agents both influence and learn from each other, thus making the overall system both unpredictable and realistic. By "realistic" we imply that this model can match rather well some of the characteristics of the real world events that are being modeled and simulated.

Of course, the modeler can go the other way around as well. She might want to define the desired goal/purpose of the system and then, with a well-defined fitness function as guidance, observe which starting conditions and agent behaviors produced the most desirable end state. The best way to define starting conditions is to utilize randomness whenever possible in order to reduce biases that so easily creep in, when we expect them the least. Although this way of modeling complex systems could be one of the most effective ways of utilizing the agent-based modeling methodology, it is rarely done in practice. This is probably due to our ingrained way of thinking that everything must be prescribed according to the perceived reality on the ground in order for the end result to be realistic or believable. Unfortunately, this attitude often results in discouraging the modeler to even start the process of simulation/modeling, mostly because of the lack of data needed to accurately represent reality with any kind of meaningful level of detail.

Regardless of the direction of the simulation approach, whether forward to the end state or backward to the initial state of the system, simulations often produce clusters of self-organizing agents. This is often manifested as synchronized movements, spatial groupings, or polarized "opinions" of agents. If we replace such clusters with higher-level agents, then we can produce a hierarchy of complex systems, where each level of the hierarchy processes information at a higher semantic or conceptual level.

Of course, in order to understand the overall behavior of the system, simulations must be run both for a long time and with many replications. This is due to the sensitivity of non-linear systems to the starting conditions. As we have already stated, random allocation of most variables is one way of mimicking nature's ways. This can be implemented both through the definition of starting conditions and the behavior of agents that learn as they move randomly about the environment.

When agents learn from each other they frequently define a community of friends, colleagues, and other acquaintances that they decide to learn from. Sometimes this community is defined randomly, while at other times it is defined purposefully, as in cases when agents decide to learn only from like-minded agents in their vicinity. This *locality* principle is critical for the success of agent-based modeling, because it promotes competition among successful candidate solutions.

Complex systems invariably exhibit non-linear behavior in some aspect of the system performance. This non-linear behavior is often characterized as a phase transition. The system "suddenly" goes from a stable state to upheaval. The same question is always asked, "Which event caused the crisis?" Most of the time the actual event is just the "straw that broke the camel's back." Any number of events could have caused the crisis, simply because the underlying irregularities in the system or the dissatisfaction of the agents have been aggregating over a long period of time. In the aftermath of the crisis it is always clear that the eruption was just a matter of time, although it was impossible to predict the actual time of the beginning of the phase transition. Examples of phase transitions include the real estate speculation-induced financial meltdown of 2008, the Arab Spring events of

2010-2013, the 1992-1995 war in Bosnia-Herzegovina, the civil rights movement in the US, and the formation of Euro zone in Europe.

In terms of statistics, such phenomena can often be characterized with power-law distributions, small-world effects, and scale-free regularities. Power laws are also known as the 80-20 (or 70-30) rule, Pareto principle, or fat tail distributions, indicating an un-even distribution of events, agents, or resources. This means that, say, 20% of the people on Earth control 80% of all natural resources. Or, 90% of users of Microsoft Word use only 10% of its features. Or, 30% of a bank's customers contribute to 70% of its revenues.

Small-world effects emphasize the greater-than-expected connectivity among all the participants in the system. This phenomenon is also known as "6 degrees of freedom," indicating that every person can reach any other person on Earth through a short chain of acquaintances.

Finally, the scale-free worlds demonstrate the presence of similar features or behavior at all levels of inquiry. For example, the Internet infrastructure looks the same, in terms of distribution of nodes and connections, at the level of the city, the county, the state, the country, and the world.

Given the dynamic, non-linear, and emergent nature of complex systems, how can their models be best used? First, they can be used to explain how or why a certain event occurred. This is equivalent to reasoning backwards from the goal state (the event that actually happened) to the starting conditions in order to obtain the agent behaviors that actually produced it. When the simulation proceeds forward, as in situations where we want to see what kind of a world will be generated from a specific set of starting conditions and/or agent behaviors, then we talk about the problem of forecasting possible futures for the system. This often leads to a "what-if" type analysis favored by many analysts and strategists. Last but not least, this type of simulation can often help uncover potential and frequently hidden risks facing an organization under various scenarios. Frequently, the greatest value of this exercise lies in the fact that it brings together decision makers and modelers tasked with improving some element of their business operations. In this way, agent-based simulations offer a common language for describing problems and devising potential solutions to these problems.

## 2.3   Examples

How do the concepts described above express themselves in different domains? One obvious example is the domain of economics. Stock markets, competition among corporations, the banking system, global trade, supply chains, internal organization of institutions, and financial crises represent examples of complex systems that we intuitively understand. All of these examples involve independent heterogeneous agents acting in their best interest within an environment that both supports and constrains their activities. Let's take an example of stock markets. In this case, agents represent individual and institutional investors who buy and sell stocks based on their analysis of the current and future values of such stocks. Sometimes this analysis includes a rigorous assessment of company fundamentals, their management resources, and their competitive landscape. Other investors use technical analysis of stock price action, which tries to identify past patterns in the price movement of a stock, to forecast future price levels. Of course, there are always people who simply act based on the advice of their favorite neighbor.

Occasionally, investors evaluate their positions and compare them to those of their neighbors, resulting in a change to their portfolio in order to capitalize on proven strategies adopted by more successful investors. As agents update their investment strategies and portfolios the stock market itself evolves, since it is just an aggregate of the behaviors of all investors. When the market changes its direction, it signals to investors that it is time to evaluate their investment decisions yet again. If the trend is downwards, the signal to investors is that they should start selling their positions. In case of a positive trend, investors either buy more stocks, decide to hold their positions, or sell stocks to realize their gains. Sometimes a strong, long-lasting upward trajectory in stock prices results in investment euphoria, leading investors to believe that the positive trend will last forever. This state of affairs was eloquently captured by the then Chairman of the Federal Reserve, Alan Greenspan, who famously mused about the "irrational exuberance" of the investors.

Predicting the movement of stock markets is inherently impossible. There are simply too many moving parts that influence each other. On the other hand, one can certainly use the complex systems framework to understand where this inherent dynamic change and volatility come from, and then adjust one's behavior to avoid being swept in a powerful turmoil brought about by the downward spiral of diminishing confidence and the well-documented aversion to losses (Kahneman and Tversky, 1984) that most investors exhibit.

Some of the Complex Systems Institute's work at the University of North Carolina at Charlotte focused on increasing our understanding of economic systems through the analogy with predator-prey relationships in ecological systems (Carmichael and Hadzikadic, 2013). This research successfully demonstrated the inevitability of the "rich getting richer" phenomenon in environments that experience infusion of resources at the lowest level of the agent hierarchy. A similar effort demonstrated the advantage of a one-price-strategy over the cascade of price levels for non-perishable goods that can be easily copied and pirated, like music and movie CDs (Khouja et al, 2008). Finally, a simulation designed by the same UNC Charlotte team, focusing on understanding the relationship between innovation and productivity improvement, demonstrated the inevitability of productivity gains possibly induced by technological advancements (Hadzikadic and Carmichael, 2011).

Another intuitive real life example is the domain of politics. Chapter 1 already introduced the DARPA-funded ACSES project which demonstrated the feasibility of formalizing and "computerizing" several social science theories to improve our understanding of how alternative actions by the US-led coalition might affect their campaign for winning the "hearts and minds" of the population in Afghanistan.

This large-scale effort captured the dynamic relationships among political, ideological, cultural, economic, and military factors in order to help analyze the viability of a handful of social science theories of allegiance in explaining the behavior of tribal leaders and their followers in specific situations as reported in the US media.

A similar effort could easily be devised for situations like the bloodshed in Bosnia and Herzegovina in the 1992-1996 period. Epstein (Epstein, 2007) has already demonstrated the feasibility of complexity science to capture the essence of rebellion and genocide. What is striking about his work is the fact that his models indicated that a complete annihilation of the "opposing" group happens in a large percentage of situations involving ethnic strife. Understanding both the causes that lead to this horrific outcome and the ways to prevent it would certainly be an effort that is well justified and badly needed.

Recent events labeled as the "Arab Spring" are but the latest example of a political situation that could use complexity science methods to forecast possible outcomes of the on-going political struggle. It is a gross understatement to say that the Arab Spring caught everyone by surprise, including the leaders of the affected countries (Tunisia, Egypt, Libya, Yemen, and Syria), leaders of the world powers (US, European Union, Russia, China), political analysts, and even the revolutionaries themselves. Who could have imagined that, after so many unsuccessful attempts to remove the ruling dictators and change the stale and corrupt regimes, the proverbial "straw that broke the camel's back" would present itself in the form of an unknown citizen who, fed up with the impotence of the Tunisian system of government, set himself on fire in front of one of the government ministries? Who could have imagined the domino effect that this action would have on the whole Arab world? Even more telling, who could have predicted that a year later voters in Egypt, when given a chance to participate in the first democratic election in Egypt in decades, would send a representative of the old regime to the run-off with a representative of the Islamists? Complexity Science has the potential to forecast such outcomes as possible futures.

## 2.4 KISS Principle: Keep It (Simple) System Stupid

Reducing Complex Adaptive Systems to "computational" practice is not easy. Chapter 3 defines and explains in greater detail the Agent-Based Modeling (ABM) paradigm that is a dominant way of implementing Complex Adaptive Systems. Remarks below refer mostly to the ABM-style implementation of CAS models.

One of the key principles of the science of complexity is that the power of complex systems comes primarily from the nature and magnitude of agent interactions, rather than from the capability of the agents themselves. Agents are often represented in a minimalistic fashion, with as few rules of behavior as possible. This is frequently the most difficult part of the system design process. After all, we suppose that we must capture all intricacies of a specific phenomenon for fear that we might miss something that will prove to be critically important in some, no matter how rare, situations. Physicians are often fond of this way of thinking. They are trained to gather all information about the patient, no matter how irrelevant it might be, before a diagnosis can be formulated.

Complex systems research is full of examples of complex phenomena being captured with a few rules of behavior prescribed to, or discovered by, system agents. Such examples can be found in the NetLogo library of models (Netlogo, 2013), where flocking, for instance, is explained with three rules of behavior (alignment, separation, and cohesion), while understanding the dynamic of AIDS epidemic requires only one rule for each participant (when in proximity, "couple" with a certain likelihood). Then, it is not surprising that simplicity is considered to be the staple of complexity.

Of course, these rules must be tested over a period of time. Given a random variation in the distribution of starting conditions, the designer must run the system many times (replications) over, for a sufficient period of time, in order to capture stochastic regularities of the system behavior. This is in addition to the need to repeat this experiment for many, if not all, combinations of the system parameters (also known as "what if" variables). Of course, all this can be automated and the system can be designed to report only "interesting" results.

This concept of interesting results must be placed in the context of the purpose of the system. Hence, defining a "fitness function" that reflects the purpose of the system is a critical step in its development. For example, if the purpose of a particular economy is to produce wealth, then the contribution of each agent will be measured in terms of its contribution to the wealth of the system. Interesting results then could be defined as behaviors of agents that contribute the most to the maximization of wealth at the level of the overall economy.

Designing a complex system is, obviously, not easy. Sometimes it is even counter-intuitive. For example, it is often non-trivial to determine the difference between an agent and an attribute. Males and females can be either two types of agent or an attribute of one type of agent called person, with "male" and "female" defined as possible values for the attribute "gender." A similar dilemma holds true for a decision regarding whether a particular concept should be defined as a rule of behavior or as an agent attribute. For example, "move north" could be a rule that an agent follows when deciding what to do next. The same behavior can be implemented as a rule that states: "consult attribute 'direction'," with north, south, west, and east being possible values for such an attribute.

Finally, designing a system as a mechanism for (often) randomly searching the space of all possible agent actions, instead of as an algorithm that anticipates all possibilities in a well-defined sequence, is often hard to grasp for system designers. It is a skill that must be carefully developed and mastered.

## 2.5   Open Issues

There are not many tools available for non-programmers to develop complex system simulations. Programming environments like NetLogo have come a long way to filling this, but they still require a reasonable knowledge of scripting languages. Visualization tools can be very useful here. However, applying such tools necessarily implies that someone else developed them in the first place, and that the end users can only apply them as is, instead of designing them according to their needs.

The question of data is a persistent issue as well. Having access to relevant data that accurately describe the population of agents, their behavior, and the environment in which they act ensures that the results of the simulation will be germane to and believable by the end user. Finding this data is not easy since, for example, many countries do not regularly capture data needed for such analyses. In addition, when this data do exist, it can be found in both structured and unstructured formats, including databases, emails, Web pages, Facebook and Twitter postings, and audio/video segments. This complicates data consolidation within one representational framework. Data capture is quickly becoming one of the most important areas for the future research in complexity science.

Verifying that the model of the system behaves according to the functional and business specifications and validating that the model truly represents the intended system and thus solves the problem is a major hurdle for the complexity science. Since these models are often designed for situations that arise infrequently, there is often not enough historical data to create a data set for model validation. Consequently, it is unreasonable to expect that the end user will use the model when its validity has not been proven. This is especially critical in situations where lives and substantial financial resources are at stake.

Finally, we should never forget that the "*customer is always right.*" End users must be a part of the team that is tasked with developing a model of a complex system. This is true for the development of any system, but it is especially critical for systems that do not have well defined verification and validation procedures.

# Chapter 3
# An Overview of Agent Based Models

Ted Carmichael

University of North Carolina, Charlotte, NC, USA

## 3.1 Introduction

It is sometimes difficult to convey to those outside of complexity studies what exactly is meant by emergence, agents, bottom-up processes, or even the phrase agent-based models itself. Part of this is due to these terms (and others) being overused. They have differing definitions across disciplines – sometimes overlapping, sometimes not – reflecting different scopes, purposes, and perspectives prevalent in these fields of study. Sometimes these terms are only vaguely defined. At other times they are defined precisely, but too narrowly to be very useful in other fields.

The word model is a case in point. Everyone already knows what a model is. The term is in widespread use, even if its meaning can change somewhat, depending on the discipline. And so, researchers might not think to question their own conception. Since everyone already knows what a model is – even if they are merely reflecting their own domain of experience – they may be slow to realize that an ABM is a different kind of model. It is different than all those – physical, mathematical, conceptual – that have come before. And so it is easy for some researchers, for example, to mistake a generative computer model for a simple, demonstrative model. Show a new ABM of, say, Lotka-Volterra phenomenon to an ecologist and he might say, "Yes, that's Lotka-Volterra. It was first proposed many decades ago. Why should I care about your version … you don't seem to have added anything?"

How do we convey what is different about an ABM, that it generates the behavior of interest rather than simply replicating it? Here is a useful analogy: imagine building a physical model of a tree, or a car, or some other known object out of Lego bricks. As with any representative model, this Lego tree is meant to be an abstraction of a real tree – a simplification that exhibits the core, aggregate properties of all (or most) trees. However, when building an agent-based model of a tree, we wouldn't concern ourselves with building the tree itself. Rather, we would build the Lego bricks. Then we would endow these Legos with ability to move, replicate them hundreds or thousands of times, throw them together in a box, and shake the box. If done correctly, the Legos will interact, self-organize, come together, and build a tree entirely on their own.

M. Hadžikadić (eds.), *Managing Complexity*,
Studies in Computational Intelligence 504,
DOI: 10.1007/978-3-642-39295-5_3, © Springer-Verlag Berlin Heidelberg 2013

This is what is meant by a generative model: one that constructs the global-level properties based only on the attributes and rules of the parts. The system is not hard-wired to generate the phenomenon of interest – only the agents themselves are so constrained, based on our best understanding of their real-world analogues. Then these agents are left to act and interact as they have been designed, so that we can see what comes of it. Only then do we return to the previous model of a tree, or Lotka-Volterra, or market outcomes, in order to validate these new models. Crucially, it is not the outcome of the ABM that is being validated, but rather the pertinent details that we have bestowed upon these constituent parts.

## 3.2 Overview of ABM

As we saw in Chapter 2, CAS is a theoretical framework useful for understanding complex systems. The main utility derives from the fact that the agents of a CAS are analogues for real-world entities (people, organizations, ants, brain cells, buyers and sellers, cars, microbes, etc.)

ABM is a natural extension of this CAS conceptual framework. The individual CAS entities, whatever they may be, are instantiated as simulated agents in the computer. Thus, there is a direct and understandable mapping from the conceptual framework to the simulated framework. As the agents in a CAS are defined locally – that is, each agent has its own rules and attributes, and makes decisions based on its own information – so are the agents in our ABM simulation. Whatever our conceptual agents need to do to participate in the global system, our simulated agents can do also. In this way, the global properties of the system of interest are allowed to arise from the aggregate behavior and interactions among the simulated agents.

ABM also has many practical advantages: primarily that non-linear *system*-level behavior is easy to represent, as the sum total of this distinct *agent*-level behavior. This does not mean that such systems are necessarily easier to program. Rather, the practical advantage has to do with our understanding of the system behavior and functionality. A direct mapping from the simulated agents to the CAS agents, and from the CAS agents to the real world, preserves an intuitive understanding of what the system represents: the mechanisms, the assumptions of the model, and the key agent-level processes.

## 3.3 Agent Based Modeling – Key Concepts

As computer simulation analogues, agents have both *attributes* (such as height, weight, color, speed, or feelings) and *behaviors* (such as actions, rules of interaction, decisions, or goals). It is also useful to distinguish these agents from the *environment* – the background in which they act. Thus, the agents can influence each other or the environment, and can be influenced in return. This influence represents the *feedbacks* in the system, such that an action at time $t$ causes a reaction at time $t+1$. Through feedbacks, both the system and the agents can adapt to other agents and the environment, allowing the global level properties of the system to emerge. This *emergence* is often the primary focus of an ABM.

## 3.4 Attributes and Behaviors

The term *agent* tends to be an overused one. Some researchers, therefore, may use an alternative, such as *particle"* (Kennedy and Eberhart, 2001) to describe the individual objects of a complex system. While logically sound in the way this is presented, it does not seem to capture the autonomy, or intent, of many agents, particularly those found in social systems. Thus we use the more conventional term *agent*, and define agents as autonomous entities possessing simple rules, attributes, and only local knowledge – information gleamed from their interactions with other, nearby agents.

We also consider agents to be largely homogenous, sharing similar attributes and rules of behavior. However, it is worth noting that many published works refer to these not as homogenous agents, but as heterogeneous agents, such as in Epstein (Epstein, 2007). We believe the discrepancy is largely a difference in semantics and emphasis. Nevertheless, for the sake of clarity, it is worth exploring the degree of homogeneity vs. heterogeneity among agents as found in a typical ABM.

Agents must, at the very least, differ spatially or temporally. Without these differences, there would be no meaningful interactions; nor would there be a way to differentiate among them. Agents in the social world are differentiated by multiple factors such as age, ideology, gender, ethnicity, economic status, educational background, or religion, and they tend to cluster in different proportions geo-spatially. So using the term "homogenous" simply indicates great similarity – even exact similarity – among the agents' rules and attributes, while it is understood that each agent represents a different current state. These agents can be identical in every way except their current state. Even if this is the case – if all the agents have the same potential set of rules and attribute ranges – they are still quite capable of producing emergent, complex behavior at the system-level. That is, even truly homogeneous agents can display correlated differences across their range of potential states, and the aggregate or global properties that result from these agents' interactions can, in the loosest sense, be considered emergent. For example, computer simulations of traffic patterns, flocking birds, or Conway's Game of Life (Gardner, 1970) exhibit various emergent properties even though the specifications for the agents are exactly the same across all cars and birds, as well as for each grid-cell in the Game of Life.

As Epstein uses the term *heterogeneous*, he is generally referring to a differentiation in terms of the agent specifications themselves; that is, a difference in their rules and/or their set of attributes. Others may characterize such agents as being *mostly* homogenous, recognizing that agents can have minor variations in terms of the rules and attributes that relate to the emergent property in question. The magnitudes of these differences across agents do matter, in their level of diversity. This is because a specific emergent property depends upon some degree of homogeneity among the rules that constitute the emergent property. For example, consider a simple traffic flow model, with the agents as cars moving along a highway. Imagine each agent having just two rules: slow down if the car ahead is too close, and speed up if it is too far away. Under some conditions, a wave-like pattern can emerge across the ebb and flow of the cars in such a simulation; as one car slows, it causes the next in line to slow, also. In simulations this can occur whether the rules for slowing down and speeding up are exactly the same across all cars, or if there is some slight variation in the threshold activation for each rule.

However, if some agents have rules that allow them to stop completely, crash, or drive off the road, then this chaotic behavior can disrupt the emergent patterns of traffic. There is a breakdown in the system at the point where agents diverge too far. In a similar manner, if the flocking example found in the NetLogo library (Wilensky, 1998) were adjusted so that some agents have wildly different attributes, then "flocking" might not be a reachable state for the system. Emergence depends upon a sufficient degree of homogeneity among the agents.

The degree to which agents must be similar depends upon the characteristics of the model being studied—specifically, only a sub-set of agent rules and attributes are generally required for a particular emergent property. For example, the agents in the traffic pattern could be made much more complex, with many more attributes than two simple rules of when to speed up and when to slow down. Consequently, agents' perceptions, dispositions, or reactive abilities could be included in the specifications. However, in terms of the emergence of waves of congestion along the highway, these attributes only matter to the degree that they relate to the two conditions that produce the emergent behavior. Therefore, even though the agents themselves may be described as quite heterogeneous, the relevant attributes must still be similar enough to produce a traffic pattern that can be analyzed and compared to real-world patterns.

## 3.5   Feedbacks and Correlated Feedbacks

Feedback, simply defined, means that the outputs of a system at time $t$ affect the inputs of that system at time $t+1$. As agents in an ABM interact, the results of some of these interactions may influence future interactions. It is this influence that represents the feedback within the system itself. In the previously mentioned model of traffic patterns along a highway, one car that slows down in response to the car in front of it may then produce a similar effect to the next car in line. This action/response that can easily produce a wave of congestion along the highway is due to feedback between the cars, from one to the next in line. It is worth pointing out that the term "wave" is apt in this case, as it describes a pattern of behavior across multiple agents, much like a wave in the ocean, even though the agents participating in the pattern change over time. This matches well with how Holland and others have described emergence in complex systems: "Emergent phenomena in generated systems are, typically, persistent patterns with changing components" (Holland, 1999, page 225).

However, simple feedback does not suffice. It is useful to consider here Warren Weaver's distinction between two different types of complexity: disorganized complexity and organized complexity (Weaver, 1948). Disorganized complexity refers to a system of many – even millions – of parts that interact at random, producing aggregate effects that can be described using probability and statistical methods. The example he gives is that of a very large, frictionless billiard table with millions of balls rolling in different directions, colliding with each other and with the walls. Even though the path of a single ball may be erratic, the system itself has measurable average properties. Clearly, there is feedback in such a system: one billiard ball strikes another, and then that ball can bounce around and strike back. But this does not suffice. There is something missing in this system that would cause the feedback amongst the billiard balls to produce self-organizing behavior.

What we are concerned with here, then, is organized complexity. Organized complexity refers to a system with a sizable number of parts that have *correlated* interactions. Furthermore, these correlated interactions produce emergent, global properties. "An average quantity alone is not an emergent feature. Yet statistical quantities which define properties of an aggregation can be regarded as simple emergent properties, if they depend on a relation of the particles to each other, i.e. if they do not make sense for a single particle" (Fromm 2005, page 08).

Note also the distinction between this organized feedback and the disorganized complexity of our billiard table. While it is true that one collision between two balls alters the course of future collisions, it does not affect the course of future collisions in a persistent way. The key point here is that such reciprocal influence among neighbors is more significant when it creates measurable, global properties. The action/reaction patterns represent the correlations within the system that make up these global properties. While our traffic pattern example may have measurable statistical properties – such as how many cars traverse the highway in a given day – these measurements do not fully capture the wave-like behavior of the system. It is by identifying the correlated feedback that we find a richer, and therefore more interesting, description of the system.

## 3.6 Defining the System: The Environment, and Endogenous vs. Exogenous Effects

One may want to consider the first action that sets a system-level pattern in motion: is it an endogenous or exogenous instigator? While the pattern is certainly endogenous to the system, the initiation of that pattern may be either. It can sometimes be difficult to characterize effects as one or the other, and how the system itself is defined may further confuse the distinction. However, by defining correlated feedback as a key property of an ABM, we bypass this argument in favor of defining what the feedback represents and what it tells us about the system.

If an external event sets off a chain reaction of persistent patterns, then the underlying properties that allow this chain reaction to occur are of distinct interest for understanding the system. If, however, there is a persistent and recognizable feedback that comes from outside of the system, then we consider this feedback to be significant in terms of our understanding of the system properties. Therefore, when we define a system, we use the method and type of feedback as a key attribute.

Consider the example of a marketplace. Such a system may encompass agents that buy and sell products or stock in companies and include the concept of wealth, earnings, and inflation. Regulatory bodies, such as the Federal Reserve, acting to tighten or loosen the terms for borrowing, may also affect the system. Clearly, if one defines the system as only the agents and how they interact with each other, then the actions of the Federal Reserve System would be exogenous to this system. However, these actions by the Federal Reserve – whatever they may be – both influence and are influenced by the state of the market. This is a significant level of feedback that should be accounted for when studying the "system," i.e., the market.

Another way of stating the idea of exogenous factors is to say that feedback goes both ways: the agents affect the environment even while the environment affects the agents. This is distinct from a model of, say, an ecology, which has sunlight as an external factor. The sun cycles through day and night, as well as annual cycles of summer and winter, and these cycles generally affect the behavior of most ecological systems. But the agents in this system cannot likewise affect the behavior of the sun. So while defining what encompasses a "system," and what potential factors are internal or external to that system, it is more important to note the level of feedback that exists between those factors, as this is both definitional and functional to the system being studied.

The type or existence of feedback suffices even with very broad definitions of "environment." If the environment for one driver-agent is defined as the road as well as all the other agents, it is the distinction between levels of feedback that is the more germane characteristic. In the models that we have developed and are described in subsequent sections, these characteristics are explicitly defined.

Emergence and Self-Organization

The term "emergence" has not yet reached a consensus definition in the literature. Some researchers distinguish between weak emergence and strong emergence, and use this definition as representing a fundamental law (Chambers, 2002).

"If there are phenomena that are *strongly* emergent [emphasis added] with respect to the domain of physics, then our conception of nature needs to be expanded to accommodate them. That is, if there are phenomena whose existence is not deducible from the facts about the exact distribution of particles and fields throughout space and [time] (along with the laws of physics), then this suggests that new fundamental laws of nature are needed to explain these phenomena."

This idea would seem to indicate that a strongly emergent property is similar to the idea of gravity: gravity is a fundamental law, a property of matter; but gravity is only apparent as one particle relates to another. In this view, it is not that the rule cannot be modeled by an agent, but rather it cannot be understood except in terms of other agents.

In our definition of emergent behavior we adopt this idea of relations among agents in the system, as in the way we have previously defined correlated interactions. A traffic "pattern" cannot really exist with only one car, and a colony of ants cannot be said to find food if there is only one ant. Thus emergent behavior is a property of a system that is at a different scale than the parts of the system (Ryan, 2007). In a similar vein, emergence is the macro-level behavior that is not defined at the macro-level, but it rather depends upon the rules and interactions of agents defined at the micro-level.

Consider a few examples of typical emergent behavior, with respect to the systems they stem from. There are the cars as agents in rush-hour traffic in the example cited previously. There are also examples of bees or ants that follow simple rules to forage for food or build a nest. Johnson talks at length about the city of Manchester, England, during the 19th century (Johnson 2002, pages 33-40). He uses Manchester to illustrate how a city with tens of thousands of people, yet absolutely no central planning, still managed to organize itself in distinct patterns, such as areas of the working class separate from the nicer middle-class neighborhoods:

"The city is complex because … it has a coherent personality, a personality that self-organizes out of millions of individual decisions, a global order built out of local interactions" (Johnson, 2002).

The brain is also often cited as a complex adaptive system, with intelligence (or even some sub-set of intelligence, such as vision) as an emergent feature. In any ABM there are often a number of emergent features, such as the self-organization of the agents and the aggregate behavior of the system. The "self" in self-organization refers to the state of an individual agent in a complex system. This agent follows its own local rules, and uses its own attributes in applying those rules.

To illustrate this idea, consider a simple model of an ant colony. For the purposes of illustration, this model need not be realistic. Assume each individual ant has the same three rules: 1) search randomly across the environment for food; 2) if you find food, carry it back to the colony and leave a "food" trail; 3) if you find a food trail, follow it until you find food.

If one ant finds food, then this new attribute – "I have food" – activates the rule to carry a piece of the food back to the colony and leave a food trail. Now, by leaving the food trail, this ant can affect the current state of any other ant that happens upon that trail. A new ant, finding the food trail, will activate its own rule to follow that trail to the food source, at which point it will also carry a piece back to the colony, and add to the trail. In this way a significant sub-set of the ant colony organizes itself to systematically collect the food and bring it back to the colony. The individual agents – in this case, the ants – are acting with limited knowledge and simple rules. But, by providing feedback to other agents and influencing them to act in similar ways, they produce the correlations of behavior that represent the organization of the overall system; i.e., the self-organization that emerges from these interactions, defining the local increase in complexity.

## 3.7   Agent-Level and System-Level Adaptation

We have previously remarked that heterogeneity among the agents of an ABM is, in the simplest possible definition, a difference in *states* among the agents. Another way of saying this is that each agent has, at any point in time, its own current value for the attributes of that agent. Thus the simplest possible definition of adaptation among agents is when their current attribute values change.

Agent level adaptation becomes hard to distinguish under certain conditions, however. In order to illustrate the potential difficulty, imagine an economic model where agents buy or sell a certain good at a certain price. The agents each have a rule that states: buy product X if it costs no more than Y units of money. Thus on one level, the agents are exactly the same, in that their internal rules are the same and they have the same *type* of attributes. But one agent's current state for the value of Y may be 10 units, while another agent may have his Y set to 11 units. The difference between them is simply a matter of variation of local conditions between the two agents. In one sense, these agents are still homogenous, because they have the same type of rules, and they apply these rules in the same way. As with spatial or temporal properties, the agents differ only in their current state for the value of Y. In another – but very real – sense, these agents are adapting individually, since the price point for each agent can vary as the simulation runs.

Agents can individually adapt on a higher level as well. For example, the rules themselves may change for individual agents, so that even if two agents are in exactly the same local situation and have the same current values for all of their attributes, they may nevertheless react to that situation differently.

The agents of an ABM also adapt in an aggregate way to their environment. This is *system-level* adaptation. An example can be found in the previous illustration of a simulation of ants finding food. We usually say that the ant colony as an aggregate entity "finds" the food, even though it is clearly individual ants that do the work. In this example the ants did not have a different set of rules or set of attribute types; they only differ in their current state of attribute values. Therefore the system is said to adapt to an environment simply because each individual agent reacts to its local conditions in a pre-determined way. These reactions do not truly represent learning among the agents; rather the agents are simply changing from one state to another.

In general, we define changes in the agents' attribute-values to be related to *system-level* adaptation – i.e., the system as a whole reacts and adapts to its environment. Thus changes in the set of agent rules themselves are *agent-level* adaptation. Much more complex behavior can result from this sort of agent-level learning – i.e., adapting or changing rules and not just attribute-values.

This sort of agent-level adaptation, this learning among agents, implies the existence of a fitness function, or selection of agents based on their level of fitness to survive or prosper in the environment. Note also that such agent-level adaptation is another argument for calling these agents heterogeneous rather than homogeneous. Without heterogeneity – without differences among the agents – then there is nothing to select for, no benefit of picking one agent as *more fit* than another. Of course, due to the actions of the agents, the environment changes continually. Consequently, the agents must constantly change their rules or attribute values in order to "fit in."

The fitness function could be defined at the system level as well. This is useful in situations where the designer wants to design a system that will have certain properties. Examples include a resilient economy, a sustainable ecosystem, a peaceful transition from one form of government to another, a nation building, or an avoidance of war. In other words, these are situations where we give the system a purpose. In such systems, the fitness of agents may be linked to the system-level function by rewarding behavior that contributes to the maximization of the system fitness function.

## 3.8  Contrasting CAS with Multi-agent Systems

As both CAS and MAS (Multi-Agent Systems) can be simulated with ABM, it is useful to contrast these two conceptual frameworks. Clearly, both CAS and MAS have many overlapping features: autonomous agents; agent interactions and feedbacks; adaptation at both the system-level and the agent-level; even emergence. However, there are also some clear distinctions that can be made, and which can further highlight the CAS approach towards modeling and under-standing systems.

Firstly, CAS agents tend to be more numerous than MAS agents: on the order of hundreds rather than a handful, or millions rather than a few hundred. Another difference is that, while agents in both frameworks are autonomous as a key feature, MAS agents tend to be more complicated and specialized. Coupled with this specialization, these agents also tend to conform more closely to a rigid hierarchy. Imagine all the parts that go into making a car. With thousands of such parts, there is ample opportunity for emergent features and feedback loops and – in particular – interactions among the agents. Yet it is also difficult to switch parts with each other. Rather, each part tends to have a specialized role. Windshield wipers don't work very well as brakes, and you can't steer the car without a steering wheel.

CAS agents are different. They are much more interchangeable. The slight variations among birds in a flock do not prevent any one of them from being replaceable with another. It is the same with ants, or brain cells, or even people in many cases. Sure, it is absolutely true that every person is unique and has his/her own contribution to make to society. Yet it is also true that, in the context of emergent global-level properties, one person, more or less, won't make much difference when it comes to sustaining a riot, determining the price of a stock, or passing along the newest internet meme.

It is for this reason that a CAS is often described with the phrase "persistent patterns, changing components." In fact the patterns themselves are generally the key attributes used for describing complex systems: "a hurricane," instead of "billions of drops of water travelling very fast in a large circle;" or "a traffic jam" instead of "hundreds of cars getting in each other's way;" or "a tree" instead of "trillions of tiny cells that want to be near each other."

The last example is rather apt, because a tree changes quite a bit over time. Yet we still might, even years later, recognize and name a *particular* tree – perhaps one that we planted in the backyard, or climbed when we were young. This same principle applies to many complex systems. The Coca-Cola Company is completely different than it was just fifty years ago: new employees, new drinks, and new distributors. But something persists, a pattern that is recognizable regardless of the particular parts that comprise it.

This interchangeability among agents is one source for the robustness of a complex adaptive system: there aren't any agents that are essential to the system. A MAS, in contrast, is much more fragile in this sense, with parts that are highly specialized and require a strict hierarchy.

In summary, compared to MAS agents, CAS agents are simpler, more numerous, more homogeneous, self-organizing, and unstructured. And while both CAS models and (some) MAS models deal with emergence, for a CAS emergence is an essential characteristic: a feature of the system.

# Chapter 4
# Running the System: Prediction and Intervention

Joseph Whitmeyer

University of North Carolina, Charlotte, NC, USA

## 4.1  Overview

Among the most valuable uses of simulation models is making predictions of how a real-world situation will develop either on its own or if interventions are carried out. Decision makers are often confronted with developing situations for which being able to predict the outcome with a sufficient degree of likelihood would be extremely useful, such as forecasting the development and course of a disease or predicting the course of an economy. In some situations, decision makers may be able to intervene in a variety of ways, which means that being able to predict the consequences of different interventions would be extremely useful. Flows of traffic or people are an example here, where simulation models might be able to predict accurately the effects of alterations to roads or buildings. For these uses, it is critical that the model be proven to be accurate through adequate calibration and testing.

Of course, the word "predict" represents a lofty goal. For simple systems like traffic control it might be even achievable. For others, like economy, ecosystem, or health system overhaul, it is not likely that we can guarantee a prescription for success. However, even if CAS can just suggest plausible futures, given the distribution of input variables and control parameters, such information would be immensely useful to the system end users.

The goal of this chapter is to present and discuss important considerations when constructing a CAS simulation model for the purpose of prediction or intervention. Many of these considerations can be summed up generally in the importance of not being inappropriately restrictive in model assumptions. Thus, the model should allow a sufficiently wide range of possibilities, should not unduly restrict possible inputs, and, in the case of interventions, should incorporate possible responses and reactions. Models should suggest likely outcomes in terms of probabilities rather than certainties. Of special interest are events that may be of low probability but of large, perhaps catastrophic, consequences. Such events are often referred to as "black swans" (Taleb, 2007).

M. Hadžikadić (eds.), *Managing Complexity*,
Studies in Computational Intelligence 504,
DOI: 10.1007/978-3-642-39295-5_4, © Springer-Verlag Berlin Heidelberg 2013

## 4.2   Key Considerations

There are four general considerations for running the system: the time horizon for which the model can be used, considerations concerning model elements, considerations concerning model outputs, and constraints on the model. The next four subsections will expand on each of these considerations.

### 4.2.1   Time Horizon

Predictions generated by a simulation model, even one that has been found to be highly accurate, are likely to be reliable only over a limited period of time. Think of models of the weather. This limitation is an especially important consideration when operating the model to explore intervention. No matter how accurately the model can predict the shorter-term effects of an intervention, the longer term effects may be much more difficult to predict. The long-term validity may even depend on the short-term result. For example, if an intervention quickly settles a turbulent situation into a stable equilibrium, that state may endure for quite some time and it is likely that the simulation model will accurately capture the long-term effects of the intervention. If, however, an intervention generates or sustains a turbulent, unstable state, then even the medium-term effects of the intervention may be unpredictable for the simulation. Our short prescription is that purveyors and users of the model should ascertain what the time horizon of reliability of the model is and should make it known to those who might make use of its results.

### 4.2.2   Changes in Model Elements

Any simulation model will have variables and constants that are built into the model. For example, in the ACSES model of population allegiance in Afghanistan that we describe in detail in a later chapter, *time elapsed* is a variable that increases at a constant rate, *people's level of security* is a variable that fluctuates according to the local situation, and the *physical layout of Afghanistan* is a constant. It is important to recognize the possibilities of uncertainty and change in these elements and to deal with these possibilities in some realistic way.

Regarding uncertainty, models may include factors, such as people's sentiment concerning some issue, fighters' combat capabilities, or the reliability of some infrastructure, for which we may have grossly imperfect information, including their initial values. In addition, changes to the system may be endogenous or exogenous and may have different levels of predictability. In a model involving a human population, the size of the population may change and individual members will age over time. In a model involving an infrastructure, the infrastructure may break down occasionally, perhaps somewhat predictably or destroyed intentionally. Exogenous events can occur as well: orders can change, attacks can come, epidemics may arrive, or borders may close. All of these events have varying levels of predictability, which exacerbates the difficulty of simulation.

With regard to endogenous changes, the modeler or user of the model has two basic choices. One is that the time horizon can be chosen to minimize the likelihood or effect of changes, in which case the changes can be ignored and do not need to be modeled. For example, although a population may be growing fairly predictably over time, if the simulation models a short enough period of time the population size may be taken as fixed. The other solution is to incorporate changes into the modeling, which we discuss below.

With regard to exogenous changes, the modeler or user faces the problems of first recognizing their possibility and, second, deciding whether they are important and likely enough that they should be included. There are two general methods of dealing with uncertainty and the probabilistic nature of many changes when making predictions. One is to make predictions contingently; the other is to build the uncertainties or changes into the simulation model somehow, and then make predictions, perhaps probabilistically. An example of a contingent prediction is: "If A's neighbor country X, seeing A's weakness, decides to attack A, then the simulation predicts an outcome with a certain probability Y. If X restrains itself from getting involved in A, then the simulation predicts some outcome with a certain probability Z." A special case of contingent prediction is to eliminate uncertainties and probabilistic changes in the model by assuming specific reasonable values for these and, then, noting these assumptions when making predictions.

Contingent predictions are particularly appropriate when investigating interventions. "If the infected are quarantined within 24 hours of the illness appearing, then the epidemic is predicted to take the following course …. If the infected are not quarantined, then the epidemic is predicted to take this course …." Contingent predictions also may be preferred when the possible uncertainties or changes come as a few, discrete alternatives, and when the modeler wants to leave estimation of their probabilities up to the user, perhaps because the user is better informed. Taking the example in the previous paragraph, the end user of the model may be in a better position to evaluate how likely it is that X will attack A. This may be sufficient enough for the simulation model to predict what will happen in each alternative.

Probabilistic predictions are generated by modeling the uncertainties or changes. Let us consider uncertainties first. If we have a single constant whose value we do not know, we may be able to estimate it and its effects probabilistically. For example, if we estimate that $x = 0$ with probability $P$ (or the proportion $P$ of the time) and that $x = 1$ with probability $1-P$, then we can perform multiple runs with the proportions $P$ and $1-P$, with values of $x = 0$ and $x = 1$, respectively.

Rather than a single constant, we may have a distribution of constants, e.g., distribution of income. Ideally, we would know the distribution empirically and would just have to assign values to agents by drawing randomly from the distribution. This might occur with age, for example, where the age distribution is usually known, even for specific subgroups such as those defined by ethnicity and gender. Sometimes we know only one statistic from the distribution, such as the arithmetic mean or the median. Often we may be able to guess the shape of the distribution as well, which will let us generate the likely distribution from which, again, we can draw randomly the values we assign to agents. Many naturally caused phenomena follow a Gaussian or Normal distribution—human height is one. Other characteristics are known to fit other distributions. For example, it is general knowledge that income and wealth usually take a Pareto or power-law distribution, with exponent between -1 and -2. Often we know the mean or median for a population we are modeling, which allows us to generate a likely distribution.

Some changes in variables may be incorporated into a simulation model in a similar way. The time between the occurrences of random events often fits an exponential distribution; the number of a particular kind of event that occurs in a given amount of time often fits a Poisson distribution. All we need, then, is an estimate for a statistic such as the mean to be able to have these events or changes occur randomly, but realistically, in the course of a model run.

A more sophisticated way to incorporate changes in a variable is to include a model of how that variable changes as a sub-model within the larger simulation model. This is likely to involve some effort, because it essentially creates another simulation model, and it may slow down the larger simulation model. For these reasons it may not be worth doing for many variables. On the other hand, for some variables a sub-model may produce the most realistic values possible for the variable, especially over time, and may improve the results and predictions of the larger model. Moreover, the user may be interested in the outcome of the sub-model in its own right. For example, in the model of population allegiance in Afghanistan, which we discuss in the next chapter, the ideology of citizens is one component of their choice of which side to support. An optional sub-model describes the change in citizens' ideology over time—and a user may be interested in the change in this variable, ideology, as well as the principal outcome of the simulation, allegiance.

With all of these techniques, the user should do multiple runs of the simulation, because assignment of values to agents will be different on each run. A caution is that the inclusion of too many sub-models or even random processes may cause volatility in results that will not correspond to reality. This can occur, for example, when the modeler assumes that sub-processes and distributions are independent when empirically they might not be.

## 4.2.3  Outcomes

One or a few outcome variables will be the focus of a simulation model created for predicting or investigating the effects of intervention. It is important in designing the model, therefore, to consider the outcome variables thoroughly so that the simulation model can answer the questions likely to be asked of it. Many of these are obvious, but it may still be worthwhile to make them explicit.

To begin with, it is important to identify at the outset the variables that are likely to be outcomes of interest, for they probably will need to be treated somewhat differently from other variables—which is what we are discussing right now. It may not be necessary to have a variable that is purely endogenous be as precisely interpretable as one that is an outcome variable. For example, if the amount of violence is an endogenous variable, perhaps a factor in stimulating population movements, what may be important is the relative difference in values for this variable in different locations rather than the actual value of the variable (i.e., specific numbers of casualties). If violence is an outcome variable to be predicted, however, users will want to know what real-world activity a given value on the variable represents.

Once the set of outcome variables is determined, probably the most crucial decision for the modeler is what set or range of values the simulation should permit for the outcome variable. Clearly, a simulation model cannot tell us how likely an outcome or consequence of intervention is if it is not among the possible values of an outcome variable. If a weather simulation does not model the production of hail, then it cannot say anything about the likelihood of a hailstorm. If a simulation model of a civil war conflict does not allow for army members to switch sides, then the model cannot estimate its likelihood. The implication here is that the modeler must take care not to unduly restrict possible outcomes, because the model not predict omitted possibilities and also it may overestimate the likelihood of permitted outcomes. The countervailing consideration is that inevitably a limit must be placed somewhere. Perhaps it is desirable to allow government troops to desert, to switch sides, to invade a neighboring country, to set up an autonomous little kingdom á la Kurz, or to become mercenaries. But at some point, if this is taken to extreme, the simulation will become unduly unwieldy in size and runtime.

Next, it is important to make sure the outcome is in the most useful form. For example, for a process occurring over a physical or geographical space, is it enough to get a value or distribution for the space as a whole, or is it preferable to obtain a spatial distribution? For an outcome that is some kind of aggregate of a variable at a lower level, such as the agent or the smallest spatial entity, is it preferable to restrict the variable to a few discrete values or to allow it to range over a continuum? Aggregation, for example, may be more straightforward with the former choice. A continuum may not only be more imprecise, but it also could present a misleading impression of accuracy if it is unclear what differences on the continuum represent.

Related to these considerations is the importance of conveying the amount of variability the model predicts for each outcome variable. If each run of the model produces a single value from a continuum for a given outcome variable, then for purposes of prediction a large number of independent runs should be carried out and both a measure of central tendency and variability, such as a 95 percent confidence interval, should be generated. For discrete-valued variables, the proportions taking different values from a large number of independent runs should be generated. It is not important whether the analysis of variability occurs within the simulation model or it is carried out by hand by the user. For purposes of predicting and analyzing intervention, understanding the variability is a crucial component.

As other chapters in this book have discussed, one of the chief virtues of agent-based models is that they allow for the possibility of generating non-linear effects. It is important, therefore, for modelers to allow for the production of these effects and for users to look for them. We discuss four here: cycles, phase transitions, catastrophes (in the mathematical sense), and black swans.

A cycle is a characteristic of a run of a simulation model in which a state of the system occurs, changes to something else, then reoccurs, all of this emerging due to the basic processes of the model such as interactions of the agents, rather than due to some kind of direction by the user or the model. Frequently, the cycle happens repeatedly, perhaps indefinitely. Some models of social unrest evince cycles: social protest increases, but police activity increases in response, which in turn produces a drop-off in protest, at which point police activity declines, which allows social protest eventually to increase again and the cycle repeats (Carmichael et al., 2010; Whitmeyer, 2009). This cyclical pattern also occurs empirically (Tarrow, 1998; Olzak and Olivier, 2005). There are two considerations worth mentioning here. First, a cycle is a pattern in time, so a user must pay attention to the course of simulation runs and not just some small set of states of the system. Second, cycles usually only occur in some portion of the parameter space, so the user must explore that space, often rather thoroughly, to uncover their occurrence and the necessary conditions for it.

A phase transition is a phenomenon in which a small change in a variable produces a large, qualitative change in a system. Mathematically, this represents the point where the partial derivative of the system state with respect to the variable in question vanishes (Poston and Stewart, 1978). There is a critical point, a value of the variable at which this occurs, in contrast to values of the variable away from the critical point where small changes in the variable only produce small, non-qualitative changes in the system. Catastrophes are a classification of more complicated patterns that involve phase transitions.

Phase transitions occur commonly in real-world multiple-entity systems, and have gotten popular attention under the label "tipping points" (Gladwell, 2000). For example, in the spread of disease there is typically a critical value of the probability of contagion above which the proportion infected increases dramatically and the disease becomes an epidemic. Agent-based models are ideal for uncovering and demonstrating phase transitions; to do so, the user must pay attention to the entire course of simulation runs and should devise a suitable metric for identifying a phase transition.

We will not go into the classification of catastrophes here, which, in any case, is unlikely to be useful for most uses of simulation models. In this chapter, we discuss only hysteresis, a common feature of catastrophes that does emerge frequently in ABMs and in real-world systems. Hysteresis refers to a situation where change in some parameters leads to a change in the system, but a return to the original setting of the parameters does not return the system to its original state. Usually something close to the original state can be obtained by adjusting the parameters back beyond the original setting; then, the original state can be regained with the return of the parameters to their original setting. Hysteresis inevitably involves the possibility of multiple equilibria, i.e., parameter settings for which more than one stable state can exist.

Hysteresis occurs in physical systems, such as the super cooling of liquids, but it also occurs in some social systems as well. For example, both low and high levels of social disorder may exist for relatively low levels of social control. A temporary large increase in social control may be necessary (Whitmeyer, 2009) to move from the high disorder equilibrium to the low disorder equilibrium.

Identifying hysteresis in a simulation model can be done in two ways. One is to change parameter settings in the course of simulation runs: from a stable state at an initial setting parameters should be changed so that the system changes, then they should be restored back to their original setting. If the original stable state is recovered, then hysteresis has not occurred; but if a different stable state now exists for the original parameter settings, then hysteresis and the existence of multiple equilibria has been found. An alternative method is to probe first for multiple equilibria by varying the initial system state for a fixed setting of the parameters. If a given setting of the parameters produces different stable states for different initial states, then multiple equilibria and almost certainly hysteresis has been found. Either method implies that the user must explore the parameter space extensively either to rule out hysteresis or to find it.

As noted earlier, a black swan (Taleb, 2007) refers to an event that has only a small chance of occurring, but if it did occur it would have significant implications. Usually if the latter is negative, the event would be devastating. Two kinds of black swans may appear in simulation models. Due to the randomness built into the model, there often is variability in the outcomes of the model for a given parameter setting and a set of initial conditions. If in a small proportion of cases there is an extreme event that can be interpreted as highly undesirable, than this is a black swan. Such an event could be the extinction or extermination of a particular class of agents, or a permanent extreme setting for some important variable in the system.

The second kind of black swan is a disastrous consequence resulting from a specific setting of the parameters, but for very nearby settings nothing of the sort occurs. It is important to note that that the disaster should not be obvious from the particular parameter setting that produces it. For example, a population model might produce vastly different results for the birth rate parameter set to 0 (i.e., extinction) and the birth rate set to a slightly positive value. This would be an expected result and would not be what we are calling a black swan.

Searching for black swans or at least being alert to their possibility means that the user of a simulation model should, again, thoroughly explore the parameter space in case there are disasters lurking for some parameter settings. Also, the user should be careful with the aggregation of results—simple reporting of central tendency and variability will not reveal, and may mask, the presence of black swans. Many simulations concern only one or a few outcomes, which could make it possible to identify a priori what would constitute a black swan and, therefore, to develop a procedure for detecting it. We mentioned extinction for the example of population changes. For businesses, it might be bankruptcy or loss of major assets. For a military simulation, it might be loss of all troops and the means for replenishing them.

## 4.2.4 Constraints

Any model, including simulation models, is at best a simplified representation of reality. The number of processes represented in the model must be limited; the number of outcomes and often their range must be limited; the number of parameters, that is, user-controlled variables that may affect the processes must be inevitably restricted as well. This situation is unavoidable. The best the modeler can do is to include the elements that theory, past research, or experience suggest are most important and to make the constraints and limitations the model imposes clear to potential users. A user of the model, in turn, should identify the constraints and limitations and should make them and their implications for any findings clear to the consumer of such findings.

## 4.3 Prediction

This section deals specifically with predictions, first in the case where a situation simply unfolds, and then in more complicated cases when, as the situation develops, other changes are likely to occur that affect it.

### 4.3.1 Advantages and Disadvantages of ABMs for Prediction

Models to predict the future are valuable in a number of areas. Examples include weather, climate, the course of a disease, movements of the economy or a segment of it, or population changes in a geographical unit or in some institution such as a school system or prison system. Several kinds of models exist, including formal mathematical models and statistical forecasting models. Agent-based simulation models can be a good alternative or complement to these, for they have several advantages. Perhaps the most important is that they impose fewer constraints on the mathematical relationships that exist between variables; they need not prescribe the relationships a priori but they can allow them to emerge. Thus, while a statistical model or mathematical model can capture nonlinearities as discussed above, such as multiple equilibria, those possibilities must be explicitly included in the model from the beginning. Any possibility for nonlinearity not explicitly included is necessarily excluded. Conversely, agent-based models allow for a variety of nonlinearities to occur simply by specifying a process at the agent level and then letting agents interact, which need not be anticipated a priori.

Another modeling advantage of ABMs is that the process at the agent level— specifying how an agent behaves—may be better understood and more accurately modeled formally than the behavior of the system as a whole. This plays to an ABM's strengths, letting the outcomes at aggregated levels emerge. One other advantage of an ABM, not to be underestimated, is that the process of how the predictions are produced is fairly transparent. Often it is even visually displayed as agents act, move, and interact. This can help a user, especially one who is not an expert, understand the predictions and how they emerged. Ultimately, it can bolster the user's confidence in the model.

It is important to be aware that ABMs have some disadvantages in prediction as well. Precisely because predicted outcomes emerge from many random events, it may be difficult to identify the exact process by which they emerged and thus to give a satisfactory explanation of them. It is more difficult to specify the effects of ABM parameters than with a formal mathematical model because the effects of the parameter space are harder to explore and understand. What outcomes tend to in the limit—as population increases, space expands, or time goes on—usually cannot be said beyond what has been simulated. When and how models can be simplified is often difficult to know. Nevertheless, many of these disadvantages are not serious weaknesses when the goal is prediction. In some cases, the exact process by which a prediction emerged may not be important as long as the predictions are accurate.

### 4.3.2  Dealing with Discrepancies

The user of a simulation model can generate predictions by specifying initial conditions, setting parameter values, running the model, and then obtaining the outcomes at some specified termination point. This is perhaps most useful for calibrating the model against data, testing the model, and showing that certain effects are possible from the theoretical perspective.

For empirical applications, however, discrepancies are likely to emerge between changes that occur in the real world and those that occur in the simulation model. The discrepancies can occur either endogenously, to variables internal to the model, or exogenously, to conditions of the model. We discuss each in turn.

*Endogenous discrepancies.* In the course of running a simulation model, some variables or constants in the model may diverge from their real-world values. For example, the model may assume that some population has some average ideological leaning, but in the real world that average leaning may shift. A model involving a social network may simulate network connectivity developing in some way that is increasingly different from how the real network is developing. These are what we call endogenous discrepancies.

There are three basic ways of dealing with endogenous discrepancies. One is to improve the sub-model of the problematic variables in the larger model—for example, replacing a constant value with a model of change. This is a good solution only if it is possible to come up with a sufficiently improved model, and if adding the improvements does not burden the model as a whole too much.

A second method is to update the variables in question in real time with empirical data. This is especially useful if the phenomenon modeled or the model exhibits chaos, such that a small initial quantitative discrepancy tends to become large and qualitative. Simulation models of the weather, for example, use this method. Its use depends on the availability of data and the feasibility of real-time feeds into the simulation model.

The third way is to generate predictions contingently. This especially makes sense if the changing variable takes a few discrete values, for example, whether a political leader chooses to contest elections or not. Then one prediction can be made contingent on the leader running and another contingent on her not running. Contingent predictions make sense if it is not feasible to update the model with empirical data—perhaps the changes are still in the future—and if incorporating a sub-model of how that variable changes will not work for some reason.

*Exogenous discrepancies.* An event can occur in the real world that seriously violates the conditions of the model and calls into question the continued ability of the model to make reasonable predictions. For example, we may have a simulation of internal conflict in a country, say a rebellion or an insurgency. Suppose, however, in the real world a neighboring country invades the country we are modeling. Or in a model of disease contagion, a mutation may occur that alters the process of disease transmission.

For exogenous discrepancies, incorporating a sub-model of change is frequently not desirable. Including the possibility of attack by a neighboring country probably would require some kind of sub-model of decision-making for the neighbor, which may be taking the main model too far afield. The same may hold for including a model of genetic mutation in a model of contagion.

This leaves two methods for dealing with exogenous discrepancies: real-time updating and predicting contingently. These warrant the same consideration as mentioned above. An additional consideration, however, is that both of these methods require the simulation model to have the ability to model the new situation. For example, the model of internal conflict would have to be able to accommodate invading troops directed against the government. The model of disease contagion would have to allow the new process of disease transmission. If these capabilities are not present in the simulation model then modelers will have to incorporate modifications or sub-models into the simulation model to handle the new elements and processes.

## 4.4   Intervention

The final section turns to intervention, which adds considerations concerning the interventions themselves, such as: noting what interventions are possible; assessing their possible effects; identifying possible reactions to the interventions, which can complicate predictions; analyzing the relationship between interventions and their effects; and assessing the cost of interventions.

Use of a simulation model to investigate the consequences of intervention is a particular case of prediction. It is predicting outcomes given that a particular change or set of changes is made to the system. Thus, the considerations for predictions discussed above hold for interventions as well. In addition, there are a number of issues that pertain directly to the interventions themselves; we discuss those in the remainder of this chapter. One cautionary remark: we make a number of distinctions regarding the kinds and techniques of intervention. These distinctions are made for convenience, as an aid in getting a handle on these matters.

One additional preliminary point is important, if not obvious. Namely, if users of a simulation model want to investigate the effects of a particular intervention, then the creators of the model must build the variables involved in this intervention into the model. For example, if users are going to investigate how lowering unemployment might affect social unrest, then employment status as an agent attribute or levels of unemployment as an environmental condition must be variables in the simulation model.

### 4.4.1   Kinds of Intervention

*Manipulation of agents.* Agent-based models usually involve a large number of agents, divided into at least a few classes. In the real world, a common way of intervening in multiple-agent situations is to manipulate the agents somehow: change their numbers, shunt them into different spaces, delay them or speed them up, and so forth. These also tend to be easy interventions to emulate in a simulation model. The model should allow agents of particular types to be added or subtracted from particular locations—so that, for example, in a simulation of conflict the user can implement a surge in troops, or in a contagion simulation the user can model the immigration of a set of vectors of a disease. Some models allow agents to be withdrawn from active participation for a period of time—put in jail, put in quarantine or simply temporarily immune, temporarily injured, and so forth. The length of this period may be manipulated in the real world and should be in the simulation model as well.

A different way to manipulate agents is to change their attributes. For example, economic security, physical security, and happiness may be important agent attributes in a particular situation, and interventions to affect these directly or indirectly may be desirable. For example, increasing terrorist acts of violence may be a permitted intervention in the model simulating terrorist networks. These acts then should be linked with agents' levels of physical security. Alternatively, the user may be given access to direct manipulation of physical security, assuming that this would help express the consequence of real world events, such as terrorist acts, not explicitly represented in the model.

*Changing the environment.* This act encompasses a divergent set of manipulations, which we can group loosely into motivations of agents, resources for agents, and behavioral options for agents. In some ABMs, agents choose behaviors based on some sort of cost benefit analysis, and so a common intervention is to alter the costs or benefits associated with particular behaviors. This would correspond to some real world policy changes, such as altering the legal penalties associated with convictions for particular crimes, changing the consequences of obtaining certain grades in school, or introducing new financial incentives for controlling the cost of healthcare. In some ABMs, agents choose behaviors following a Markov-type model, such that different behaviors are chosen with set probabilities or drawn from a probability distribution. Alteration of costs and benefits can be accomplished implicitly here by changing the corresponding probabilities appropriately. Thus, a relative increase in the rewards for going to college, say, could be implemented by increasing the probability a student agent enters the college-going state.

A common kind of intervention in the real world is to change resources. People donate money to politicians and charitable organizations to help them achieve their goals. Countries give aid of various kinds to other countries. A way of combating rogue states, organizations, and people is to cut off their access to resources, such as guns, money, and recruits. If agent resources are built into an ABM, then the ability to alter agents' resources is a fairly easy and logical intervention to implement.

### 4.4.2   Techniques of Intervention

Intervention can be timed in several different ways, depending on the goals of the user. One way is to run the simulation program with different initial conditions, with the assumption that an intervention has already occurred to create those conditions. For example, the user could run the simulation with different starting amounts for one kind of agent—say, police, assuming an intervention of putting a certain quantity of police on the ground. The same results would be achieved by adding those agents in the course of the simulation run, if it is early enough that other initial conditions have not changed. The advantage of varying the initial conditions is that the values of other variables at the time of intervention are controlled completely.

The user can intervene in the course of a simulation run either according to the time elapsed or in response to some situation observed in the run. In either case, for the user to make claims about the effect (including no effect) of intervention, a systematic testing of results should occur, including exploration of alternative timings, different levels of intervention, and multiple runs for each policy (see our fuller discussion of analyzing intervention, below).

"Response to some situation observed in the simulation run" corresponds to many real world cases where an event occurs which forces an entity such as a government to intervene. Note that intervention may be more effective if it occurs earlier, in response to change in a variable that tends to be an early warning for the event in question. For example, an outbreak of social unrest may tend to be preceded by a drop in some attribute of people agents such as happiness, or by the release of a large number of people agents from imprisonment, or by a conjunction of these. It may be worth exploring whether intervention—say, increasing the number of police—in response to the changes in these early warning variables, rather than to the subsequent event of concern—say, the outbreak of unrest, might be more effective. Exploration of this may require modification of the simulation so that possible early warning variables are displayed in such a manner that the user can respond to them with intervention. Alternatively, this exploration can be automated. Results of these investigations could be a valuable input to policy decisions in the real world.

### 4.4.3 Incorporating Responses to Intervention

In the real world, intervention is likely to be met with a response. Some responses may emerge from the simulation model, but this capability has to be built into the model. The modeler, therefore, must also anticipate what possible responses to a particular kind of intervention might be, and make sure those responses are possible. If not, it is probably advisable to add their possibility to the simulation model. This is an extremely important aspect of creating simulation models. Unanticipated responses and consequences are a continual hazard for actual interventions and one strong advantage of a useful simulation model is that it can help users identify such responses and figure out how to deal with them.

For example, in a simulation of social unrest, increased repressive activity may meet with the following responses: 1) decreased dissident activity, 2) a shift toward underground forms of dissidence, 3) increased radicalization in the population, and 4) population flight. These may be alternatives or may occur in some combination. Other responses may exist, as well, such as 5) foreign intervention against the repression, in the form of increased resources for the dissidents or even military action. Omission of possible responses may be possible for some responses—for example, it may be not be important for the goals of the model to distinguish between overt and covert dissidence, in which case response 2) above would be moot. Even response 4) may be ignorable if population movements are not important. Assuming that the purpose of the simulation is to gauge the effectiveness of repression and how it varies across differing responses, none of the other three responses should be ignored.

The difficulty for the modeler is that it may be impossible to anticipate all reasonably possible responses to an intervention. In the above example, the fifth response, foreign intervention, depends on looking outside the simulation environment and considering a potentially large set of countries and their interventionist proclivities. Because of this problem, it probably will be beneficial to consider empirical research on the situation being modeled while creating the model. The research literature and experts may identify responses that the model must incorporate to have value and credibility. Common sense may identify other plausible responses as well. Finally, a non-prescriptive model of the agents in an environment that is not overly controlling may still allow other responses to emerge.

## *4.4.4   Analyzing Effects of Intervention*

The goal of investigating intervention through a simulation model is to analyze its effects. How this is done will depend on what is desired from the analysis. Initially, the user is likely to explore outcomes of the model through a process of trying interventions, observing the results, and repeating the intervention while making adjustments to initial conditions, parameters, or timing of interventions. Such exploration is usually desirable at least for the user to get a feel for how the simulation proceeds. For some purposes the results of this process may be sufficient—if the goal is, say, to identify a range or set of possible effects of interventions.

In many cases, however, the user will want a more thorough and exact understanding of the model's predictions of the consequences of intervention. To begin with, the user may want to know not only the set of results of intervention but their probabilities as well. The user is likely to want to know how variations in application of the intervention affect results—variations in amount and timing, at a minimum. It may be that initial conditions and parameter settings are given, but some sensitivity analysis will greatly enhance understanding of any resulting effects of intervention. That is, the user should investigate the extent to which small variation in initial conditions and parameter settings affects the outcomes.

If the type of intervention is not prescribed, or if initial conditions are not specified, or if initial parameter settings are not preset, then the user should carry out a systematic survey of the range of possible values for all parameters, variables, conditions, or intervention types. For a given configuration of settings and interventions within this survey, the user may want to perform repeated runs in order to obtain information on relevant probabilities. Moreover, for any component that is not being surveyed—say, initial conditions—a sensitivity analysis should be performed. Often the user will want to use the surveyed space to identify necessary and sufficient conditions for producing certain key outcomes. In this regard, we note again that it may be important to look for black swans— events of low probability but of great importance, usually in a negative sense.

Much of this analysis can be automated, regardless of whether the issue is a systematic survey of the effects of parameters, initial conditions, interventions, or a sensitivity analysis around some particular configuration. However, the user should be judicious in this step. For example, establishing an appropriate range of values and an appropriate number of repetitions for each configuration value may vary depending on the results. Around critical points such as phase transitions, where the outcomes seem to change qualitatively or perhaps where certain outcomes start to become possible, the user may want to adjust the automated survey to hone in more finely on that particular region of the configuration space. The search for key outcomes such as black swans can be automated as well.

One final remark is that for some investigations of intervention it may desirable to also include cost. The user may want to know where one gets the biggest "bang for the buck." The simplest way to estimate cost would be to include cost estimates for each kind of intervention. More elaborate methods are possible as well. The results can be assessed in several ways, for example, as a simple average, success per unit cost, or perhaps a graphical display of the relationship between cost and best outcomes.

## 4.5  Conclusion

In this chapter, we have presented key issues involved in the two principal kinds of applications of simulation models: predictions and interventions, a specialized form of predictions. We have discussed timing, inputs, and outputs, the three basic elements of predictive models. We went over some of the challenges for using ABMs for predictions, and surveyed the techniques and issues involved in using simulation models to investigate interventions.

In order to use simulation models for prediction and for analyzing interventions, it is often helpful to maintain awareness of work that is being done in related areas. We mentioned above that it is useful to know empirical findings regarding the modeled phenomenon as it occurs in the real world—findings on the historical course of actual disease, for example, if one is modeling epidemics. Here we note two other kinds of research on similar phenomena that modelers should consider: other simulation models and formal mathematical models.

Other simulation models are likely to take different approaches, even the ones that also are ABMs. They are likely to make different assumptions. Comparisons of the models will help in the clarification and justification of one's model. Comparisons of the results will help reveal what findings of the model are robust, what results are new and surprising, and perhaps what outcomes might be suspect and need further investigation.

Comparing simulation models to mathematical models can have some of the same benefits. It can tell modelers whether their models fit what one might expect from mathematical models. It can suggest where the modelers might need to probe more deeply, for example, if they have not found a critical point or phase transition where one exists. Comparison with mathematical models may also suggest where and how one may uncover regularities, such as limiting distributions once one has normalized outcomes appropriately (see any textbook on stochastic processes, e.g., Karlin and Taylor, 1981).

Finally, we note that much of this chapter has been at a general level, of necessity because we wanted to present the ideas and issues that will be important in many specific cases of prediction and intervention. Many of these ideas and issues, however, can be seen more concretely in our description of the ACSES model of Afghanistan.

# Chapter 5
# The ACSES Model of Afghanistan: Introduction and Social Theories

Joseph Whitmeyer

University of North Carolina, Charlotte, NC, USA

## 5.1  Introduction

In this chapter we present an example of the principles and approaches we have described to this point. The example in question is a model of population allegiance in Afghanistan in the face of the Taliban insurgency, and effort by the Afghan government and coalition forces to combat the insurgency. DARPA supported our development of this model as a pilot project, which we refer to it as the ACSES ("Actionable Capability for Social and Economic Systems") model.

The ACSES model is an agent-based computer simulation model run on a Netlogo platform. The purpose in building this model was to demonstrate that a useful ABM model could be built that would: 1) explain or predict a population outcome of interest to policy-makers and strategists, 2) be able to make use of data of various kinds, 3) be testable with real-world data and perform well at prediction, and 4) implement different social theories so that the efficacy of the theories could be tested.

The population outcome of principal interest in the ACSES model is citizen allegiance, i.e., the number of citizens supporting the Taliban, supporting the government, or being neutral. The simulation takes place on a map of the physical space of Afghanistan, which is realized in the form of a grid consisting of 835 square patches. Accordingly, the model generates the geographical distribution of citizen allegiance.

The principal agents in the model are of four kinds: Taliban insurgents, coalition forces, government forces, and ordinary members of the population, whom we call citizens. The simulation also implicitly involves leaders of various ethnic groups in that the user specifies the allegiance-related orders and some characteristics of the leaders, which then affect the citizens' allegiance. The leaders, however, are not represented visibly in the simulation display.

The model incorporates data concerning the distribution of population over the land, the geographical distribution of ethnic groups, and the economic situation of the population, again varying according to location. The model also gives the

M. Hadžikadić (eds.), *Managing Complexity*,
Studies in Computational Intelligence 504,
DOI: 10.1007/978-3-642-39295-5_5, © Springer-Verlag Berlin Heidelberg 2013

option of calibration and testing of the model by comparison with actual data concerning violent events in one district of Afghanistan over a three-year span in the 2000s. Calibration of the model consists principally of choosing the social theory that does best at reproducing the empirical pattern of violence. The model incorporates social theories in such a way as to allow a continuum of different possibilities. To facilitate our analyses, however, we specified 12 distinct theories. In addition, we built into the model four theories of how actors, over time, may change in characteristics important for the social theories.

In operation, the model works as follows. Prior to running the model, the user must choose the social theory to be operative, whether the run will make use of the violence data (and, therefore, be a calibration or testing run), and any deviations from the default settings for other parameters, including the leader attributes of resources and ideology and citizen ideologies in various regions. Once the user has started the simulation, the user may make certain changes as the simulation runs to incorporate events that may occur. The important changes include the addition of any of the three kinds of forces to any specific region, thus modeling a troop surge, changes in leader resources, and alterations of leader ideology, which describes a change in a leader. In the real world, this could be due to assassination, imprisonment, etc. While running, the simulation allows the user to switch between views of the distribution of agents over the map or distribution of allegiance, and also displays ever-lengthening graphs of the time history of the numbers of the various kinds of agents.

### 5.1.1  The Afghanistan Setting

Afghanistan was chosen as the setting for the ACSES model because 1) it exhibited the general features of the situation we wanted to model; 2) it was a situation of interest and importance at the time; and 3) some data were available on the situation, both for informing the model and for calibrating and testing it. The political and military picture of Afghanistan in the second half of the 2000s was that an Afghan government supported by a U.S.-led coalition was in control, to varying degrees, of the country, but was combating an insurgency by the Taliban.

Afghanistan was and is one of the poorest and least developed countries of the world, with a history of undergoing foreign interventions that have encountered serious difficulties, in part due to the rugged landscape. In the immediate past, in the wake of the 9/11 attacks on the U.S., a U.S.-led coalition and regional strongmen had replaced the Taliban-controlled government and had pushed the Taliban out to remote areas of the country and across the border to Pakistan. Despite democratic elections and installation of a pro-Western Afghan government, the Taliban had regained some of their former strength and had mounted potent insurgencies in several parts of the country. Afghanistan has a number of distinct ethnic groups, located for the most part in different parts of the county. The most numerous ethnic group is the Pashtun, comprising over half of the population. The Taliban are primarily Pashtun and their appeal is primarily to the Pashtun; the insurgencies are almost entirely in Pashtun regions.

Citizen allegiance was a suitable outcome to model for Afghanistan because, first, it is likely to be important both for short-term military struggles and for the long-term political and social future of the country. Second, it was a characteristic that was very much in play at the time. For example, former Afghan Interior Minister Ali A. Jalali in a 2007 interview claimed: "Most surveys claim that only 10 percent of the people are fighting for [the] Taliban and 20 percent are fighting for the government, while 70 percent are sitting on the fence" (Bhatt 2008). The situation in Afghanistan also was suited to simulating the effects of various interventions because both sides were varying their deployments and tactics (Landay, 2007; St. Onge, 2006).

Typical of poor and less developed countries, some data were available for Afghanistan on important items such as the size and distribution of the population, distinguished by ethnicity, the ideological leanings of different ethnic groups, and the economic condition of people in different regions. We also had time history data on violent events over a three-year period for the Ghazni region, which we were able to use to calibrate and test the model. This availability of some data but its incompleteness is a typical state of affairs for many situations it would be desirable to model. This also made Afghanistan a good setting for a pilot study to demonstrate feasibility.

## 5.1.2   Goals of the Project

The general goals of the project were 1) to demonstrate the feasibility and potential usefulness to an end user, such as a commander in the field, of a simulation model of citizen allegiance in a region experiencing an insurgency; and 2) to demonstrate the possibility of a simulation laboratory for social theories, which would allow implementation of numerous alternative social theories and comparison of their results.

The final ACSES model, especially as applied to the struggle in the Ghazni region of Afghanistan from 2005 to 2008, shows that the simulation may be useful in explaining or predicting citizens' allegiances and some consequences in situations of insurgency. Thus, if the Taliban mount an offensive in a region of Afghanistan, the simulation may help a decision maker answer the following questions:

1) How many citizens in the region initially support the Taliban?
2) How are the allegiance of citizens, their ideology, and their support for the different factions likely to change over time?
3) What will be the violence level in the region if the coalition forces fight?
4) How would gaining the support of local leaders for the coalition forces change the answers to questions 1-3?

5)  How would a shift in the support of local leaders to the Taliban change the
    answers to questions 1-3?
6)  How would sending additional coalition forces into the region change the
    answers to questions 1-3?
7)  How would increasing or decreasing the resources of local leaders affect the
    allegiance, ideology, and behavior of citizens?

The final ACSES model can be used to test the effectiveness—i.e., the ability to
explain and predict—of a variety of social science theories. A range of theories is
possible at three different levels: individual action, social effects of aggregating
individuals, and psychological change within individuals. The model allows
adjustments to the theories and using a combination of the theories, both within
and across levels. In addition, while the range of theories that can be used is
necessarily limited, the mechanism for implementing the theories allows for easy
expansion. The model allows the user to answer questions such as:

1)  Do different social theories produce qualitatively different results (e.g., the
    answers to questions 1-7 above)?
2)  Do theories at different levels combine to produce new results?
3)  Are the effects of combining theories, whether within or across levels, linear
    or nonlinear?

## 5.2    Agents

In this section we describe the five kinds of agents involved in the ACSES model.
Four of these—citizens, Afghan government fighters, coalition forces fighters, and
Taliban fighters—are instantiated as visible agents in the simulation. The fifth,
leaders, is present only through their effects on other agents.

### 5.2.1    Citizens

Citizens are by far the most numerous agents in the simulation model. The citizen
agents' allocation of allegiance—to Taliban (-1), to the Afghan government (1), or
neutrally, to neither (0)—produces the outcome of principal interest, i.e., some
aggregation of citizen allegiance. The allocation of allegiance is the behavior that
is modeled in the simulation by the social theories as realized via the utility
function. We describe the utility function in detail in a subsequent section.

   Allocation of allegiance is, in fact, citizen agents' only behavior. For example,
they cannot move from one patch to another. There are a few implicit behaviors of
citizens. Because the allegiance of citizens on a specific patch affects the outcome
of combat on that patch, they implicitly may act to help one side or the other in
conflict. If theories of social influence are activated, they implicitly may act to
get nearby citizens to act similarly, say, through social pressure, conveying
information, or example.

Citizens have several attributes. One is in fact the allocation of allegiance they have chosen: a citizen can be a neutral (0), a Taliban helper (-1), or a soldier helper (+1). This is important for the course of the simulation because it affects combat between fighters and it can affect other citizens' allocation of allegiance if social influence theories are being used.

Two attributes are fixed for an agent: the patch on which the agent is located— because agents cannot move to a different patch—and the ethnic group to which the agent belongs, which corresponds to the region in which the agent's patch lies. The citizen agent's perception of the legitimacy of the agent's leader is an attribute relevant only if the theory of psychological change with respect to ideology is activated. The remaining attributes correspond to the seven elements of the utility function: values indicating the citizen agent's economic situation, security situation, amount of coercion faced, ideology, loyalty to the leader of the agent's group, amount and direction of social influence experienced, and amount of resistance to repression that is occurring near the agent. We present the utility function and theories of psychological change later in this chapter.

## 5.2.2  Fighters

Taliban, Afghan government soldiers, and coalition forces soldiers are the three kinds of fighter agents implemented in the ACSES model. Each time step of the model, all fighters of a given kind on a given patch carry out one of four behaviors: they fight, make an organized move, stay on the patch without fighting, or flee. Which behavior the fighters choose depends only on the situation in their patch. It is influenced by the strength of their side on the patch, their perception of the strength of the opposing forces, and by the degree of support their side has from the local population. In essence, if fighters think things look bad for them, they choose to flee; if they think things look good, they choose to fight; if they think their situation is somewhere in the middle, they choose to move or to stay. We describe this choice in detail here, without, however, the numerous equations that implement it.

Two attributes of fighter agents are crucial to their behavioral choices: their effectiveness at fighting and how they perceive the other agents on their patch. A value for effectiveness is assigned to all fighters of a given kind; it is intended to correspond to the quality of their training, command and control, weapons, and so forth. Agent perceptions constitute a large set of parameters, for they are indexed by three variables: the category of the perceiver, the actual category of the agent perceived, and the category the perceivers perceive. There are six relevant categories: coalition soldier, government soldier, Taliban, soldier helper, Taliban helper, and neutral citizen. We assume agents accurately recognize others in their own category, for they are likely to know them personally. That still leaves five categories to perceive and five categories to perceive them as. So, for example, coalition soldiers might accurately perceive 5 government soldiers as 5 government soldiers, but 10 Taliban as 15 Taliban, 10 Taliban helpers as 5 Taliban and 5

Taliban helpers, 10 Afghan government fighters as 8 government fighters and 2 Taliban, and so forth.

Out of the perceived numbers of different kinds of agents, the perceived effectiveness of opposing forces, and the actual effectiveness of their own forces, each kind of fighter estimates its relative advantage or disadvantage on the patch and makes its behavioral decision accordingly, based on threshold levels set by the user in the initial setup. Afghan government soldiers and coalition soldiers choose the same action. Thus, although they have different abilities to identify agents and different effectiveness, these variables are averaged according to the relative proportions of the two kinds of fighters in the process of determining the behavioral choice. All fighters follow their preferred behavior except that if one side chooses to fight when the other side has chosen to move or stay, the latter side is forced to fight as well.

If fighters choose the "move or stay" option, then there is a further choice: to stay in the patch or to move to a neighboring patch. The choice between these two is made randomly according to a weight specified by the user in the initial setup. If "move" is the choice, all fighters of the given kind in the patch move to the neighboring patch. For the "move or stay" choice and for the choice of destination patch, coalition soldiers and government soldiers are no longer combined. "Flee" is also a move to a neighboring patch chosen at random, but all agents of the kind in question do not move as a coherent mass to the same patch, but rather each agent randomly chooses its destination patch independently. In this way, "flee" differs from "move."

Finally, combat produces casualties according to success likelihoods specified by the user in the initial setup but then weighted by the relative numbers of agents on each side in the given patch. There is a further negative adjustment to the success likelihood for a side that chooses not to fight when the other side chooses to fight.

### 5.2.3  Leaders

As noted previously, leaders are not implemented as actual agents in the ACSES simulation. Instead, the leadership for each ethnic group except for the Pashtun is implemented as two variables that represent conditions for the citizen agents of that ethnicity. The Pashtun, as the largest ethnic group, distributed over most of the country, are assigned six separate leaderships according to region. The two variables are the leadership orders—to support the government, remain neutral, or support the Taliban—and leadership resources, which under some social theories affect how much influence the leadership orders have on the allegiance decisions of the citizen agents. The section on the utility function details how these variables affect citizen allegiance.

The user sets the values of leadership orders and resources initially but also can change them as the simulation is running. A change in orders could represent, for example, the leadership changing its preference for some reason or the leadership being supplanted or replaced by a leadership with a different position. A change in resources might be due to actions by the government, local actors, the Taliban, or other factors, or, again, be due to replacement of the leadership.

## 5.3   The Implementation of Social Theories

Because the focus of the ACSES model is on the behavior of the citizen agents—specifically, their allocation of allegiance—its most sophisticated models are of these agents. The ACSES simulation includes models of the citizen agents at two levels and time scales. One model is at the behavioral level, that is, yields behavioral choices, and instantiates social theories concerning allegiance. The other model is internal to agents, that is, changes parameters of the behavioral model, and instantiates theories concerning psychological change. The effects of the psychological change model occur much more slowly than those of the behavioral model.

The theories of allegiance are implemented via a utility function for the citizen agents. In this section we describe the role of the utility function in the simulation model, the basic operation and characteristics of the utility function, and why this form of a utility function was chosen. We then explain the particular allegiance theories that were implemented in the ACSES model, and how the utility function was used to implement them. Last, we describe the psychological change models.

One side remark is that, as described above, the ACSES model includes behavior by fighter agents in addition to the allocation of allegiance by citizen agents. The model of the fighters' behavior is less sophisticated, however. It is simply a Markov decision model: a variable value is calculated to represent the fighter's evaluation of his current situation and the cutoff points in the possible range for the variable dictate which behavior he chooses.

### 5.3.1   The Utility Function

The ACSES model uses a utility function to model citizen agents' allocation of allegiance. While many different utility functions exist, the basic modeling process is the same: for each considered behavior, the utility function combines the states of the agent on any number of variables into a single quantity, labeled "utility." The agent then chooses the behavior with the greatest utility. Social science theory and empirical findings suggest that the choice of allegiance in conflict situations such as that in Afghanistan is a complicated process that involves many variables. A utility function model is one of the best ways to model such decisions because on the one hand it can instantiate numerous theories and quite complicated ones, and on the other hand it produces, via numerical calculation, a single behavioral outcome, which is necessary for an agent in a simulation model. A utility function model can incorporate all variables that theories deem to be relevant, it can embody hypothesized relationships between variables, and it can instantiate different theories by adjusting parameters.

Many utility functions have been devised. The ACSES model uses the Cobb-Douglas utility function, which specifies utility as the product of preferences (i.e., values, interests, motivations) raised to a fractional power. Equation (1) shows the general form of a Cobb-Douglas function, with $P_i$ denoting the $i$-$th$ out of $n$ preferences, and $w_i$ indicating the relative weight or importance of that preference:

$$U = P_1^{w_1} P_2^{w_2} P_3^{w_3} ... P_n^{w_n} .$$ (1)

The exponents are required to sum to 1.

The simulation model uses the utility function by calculating, at a given moment, for a given agent, the expected utility of each behavior in a set of considered behaviors. The agent, then, chooses to perform the behavior, from that set, that has the highest expected utility. The Cobb-Douglas utility function has been used for instantiating a variety of social theories through both formal and simulation modeling (Coleman, 1990). It has several desirable features, including simplicity and modularity, which means that it is easily expandable to include additional preferences or values or motivations if they are important for a theory. The incorporation of a large number of preferences in the utility function does not compromise its simplicity, because any preference $P_i$ can be turned off by setting its weight, $w_i$, to 0. Thus, if they so desire, users can operate with what is effectively a much smaller utility function. The Cobb-Douglas function has the additional desirable feature that losses have a larger effect than gains, which fits experimental findings (Tversky and Kahneman, 1981). Mathematically, this means that the second partial derivative of the utility, U, with respect to any given value is negative. Before deciding on the Cobb-Douglas function, we also tested a least squares utility function, which had many of the same desirable features. We preferred the Cobb-Douglas function, however, because it showed greater sensitivity to changes in parameters.

## 5.3.2   The Behavioral Theories

Through the utility function, the ACSES model instantiates social theories of allegiance involving obedience to leaders and involving social influence; it also incorporates general relevant characteristics of the agent and the agent's circumstances. All of the theories have substantial support in social science literature as being important. We briefly reference some of this literature here, but in order to remain close to the objective of this chapter, which is the presentation of the ACSES model, we do not otherwise discuss the social science.

The ACSES model instantiates three specific theories concerning the obedience of citizen agents to a leader's orders. These theories are relevant in situations in which—see above about leaders—leaders, with given levels of resources, order their followers to adopt certain allegiances. The legitimacy theory posits that a citizen agent wants to follow the orders of the leadership simply because it is the legitimate leadership. This idea is common in the social science literature. It corresponds to the concept of authority either granted willingly by an agent, or delegated or enforced (Coleman, 1990). It fits the idea of legitimacy granted to group members who have certain characteristics or demonstrate certain abilities (Berger et al., 1998). It also can encompass Weber's (1968 [1922]) concept of charisma, according to which some individuals have qualities that induce people to follow them. Legitimacy is realized in the utility function by giving a high weight, $w_L$, to the preference variable loyalty, given by $(1\text{-}L)$. As with all preference variables, the relationship here is defined as an equation $0 \leq (1\text{-}L) \leq 1$. L is given by the difference between the leadership's order and the agent's considered behavior, all weighted by the legitimacy of the leadership. Thus, $(1\text{-}L)$ is greater—i.e., closer to 1—the more the agent's allegiance coincides with the leader's order.

The coercion theory posits that a citizen agent wants to follow the orders of the leaders because he or she believes they are likely to punish those who do not obey or to reward those who do. This theory can be traced back at least to Machiavelli, who advised in The Prince (Machiavelli, 1985 [1513]) that rulers could be most effective by making their subjects fear them, and also has contemporary advocates (e.g., Kiser and Linton, 2002; Levi, 1988). It fits the behaviorist perspective in psychology (see, e.g., Sidman, 2000) and deterrence theory in criminology (Becker, 1968; Stigler, 1970). Coercion is realized in the utility function by giving a high weight, $w_C$, to the preference variable coercion, given by $(1\text{-}C)$. Understanding that $0 \leq (1\text{-}C) \leq 1$, the agent model has that $(1\text{-}C)$ deviates from 1 more the more the agent's allegiance deviates from the leader's order, weighted by the relative resources the leader can bring to bear on the agent.

The representative theory posits that a citizen agent follows the leadership only if the leadership advocates what the citizen agent wants. Thus, the leadership's orders have only a weak effect on the behaviors of the population. This idea goes back at least to Marx (1990 [1867]), who saw the government as merely implementing the desires of the ruling class, but it also has been argued for recently, for example, in the context of Latin American rebellions and popular nationalism (Whitmeyer, 2002; Wickham-Crowley, 1991). To realize the representative theory, the weights for loyalty $(w_L)$, and coercion $(w_C)$ are both set low, with the weights for the general variables (see below) set relatively high.

Three general variables likely to affect a citizen's decision, no matter which obedience theory may apply, are the citizen's ideology, her economic situation, and her security situation. Thus, in the ACSES model, a citizen agent may be affected by:

- An ideology predisposing her to a given allegiance position;
- Her economic situation and her understanding of how different regimes might affect it; and
- Her expectation for the amount of security from violence offered by different regimes.

In the utility function, the variable $(1\text{-}I)$ measures the conformity of an agent's projected behavior with her ideology; $(1\text{-}E)$ measures the closeness of an agent's expected economic situation under a given projected behavior to the optimal economic situation; and *(1-V)* indicates the level of security from violence the agent expects under a given projected behavior. To instantiate the representative theory the weights, $w_i$, $w_E$, and $w_V$ for these three variables are set relatively high. The weights are still nonzero, but much lower, when instantiating the other obedience theories.

We turn next to social influence. There is considerable empirical evidence that people's participation in collective action is affected by the behavior of those around them and those close to them. This is true around the world, for example in the United States (e.g., Roscigno and Danaher, 2004) in the former East Germany (Opp and Gern, 1993) and in late twentieth century China (Calhoun, 1994). Many theoretical models of collective action exist, and most if not all of them incorporate the effects of social influence (Oliver, 1993; Ostrom, 2007). It would be a mistake, therefore, to neglect social influence theories in modeling the allegiance of the population in a country with significant internal conflict.

In the ACSES model, the utility function for citizen agents instantiates two different theories that concern the effects of social influence on participation in collective action (see Table 5.1). The first social influence theory stems from research, in part based on the study of riots, that shows that a person is more likely to join collective action the more his or her immediate friends do (McPhail, 1991, 1994). Because the ACSES model does not incorporate a detailed representation of friendship networks, we approximated this effect with the assumption that a citizen agent may be influenced to do what similar citizen agents near him do. We labeled this theory simply social influence, although contagion might be a more precise descriptor. Social influence is realized in the utility function by giving a positive weight, $w_F$, to the preference variable social influence of close associates, given by *(1-F)*, with $0 \le (1-F) \le 1$. The term *(1-F)* is greater—i.e., closer to 1— the more the agent's allegiance coincides with allegiance of those around her, weighted so that the greater the proximity to the agent the more effect on *(1-F)*. A second parameter sets the radius of social influence affecting a single citizen agent.

The other influence theory—labeled resistance to repression, or just repression—encompasses two countervailing effects. First, people are less likely to join a collective action the greater the local presence of forces to repress it. Second, in the presence of some repression, people are more likely to join collective action the larger the proportion of people in the observable population who are engaged in this action. These effects are shown, for example, by studies of the overthrow of the East German regime in 1989 (Opp and Roehl, 1990; Opp and Gern, 1993). They are also present in the critical mass idea that often collective action can accelerate once a critical mass of people has joined (Marwell and Oliver, 1993).

Resistance to repression is realized in the utility function by giving a positive weight, $w_R$, to the preference variable *(1-R)*. Repression is a consideration (i.e., $R > 0$) only for citizens who are considering helping the side that is outnumbered on their patch. If the considered behavior is 'neutrality' or 'helping the predominant force on their patch,' then no repression occurs ($R = 0$). If repression occurs, it is less effective—i.e., *(1-R)* is closer to 1—in the presence of an increasing number of proximate citizens who are defying repression. A second parameter sets the radius of the area a citizen agent observes when evaluating how prevalent a pro-government or pro-insurgent behavior is. The effect of resistance is, therefore, that citizens are less likely to defy factions that have a relatively strong presence on their patch, especially when that faction has little support among other citizens in the general area.

The option of the social influence theories is likely to be useful to a user of the simulation model who can employ them in conjunction with one of the obedience theories. Neither the influence of those nearby nor resistance to repression is likely to initiate pro-government or pro-insurgent activity. For that purpose some other cause, such as a leader's order, is needed. However, the repression theory adds the ability of local forces to suppress unwanted activity. Both theories bring in people's influence on each other, which introduces the possibility of activities spreading across a population. These features will increase a user's ability to use the model to assess the efficacy of measures and countermeasures. We remark also that social influence may mitigate or exacerbate other effects of people's situation and, as research has shown, is likely to be responsible for nonlinear and emergent effects in social phenomena generally (Coleman, 1990) and in collective action in particular (Ostrom, 2007).

Equation (2) gives the specific version of the Cobb-Douglas utility function (see equation (1)) that is currently implemented for citizen agents in the simulation model:

$$U = (1-L)^{w_L}(1-C)^{w_C}(1-I)^{w_I}(1-E)^{w_E}(1-V)^{w_V}(1-F)^{w_F}(1-R)^{w_R}. \quad (2)$$

Each citizen agent calculates her utility for three different behaviors—supporting the Taliban, supporting the government, or being neutral—and chooses the behavior with the highest utility. $L$ (loyalty to leader), $C$ (coercion), $I$ (ideology), $E$ (economic welfare), $V$ (security against violence), $F$ (influence of close associates), and $R$ (repression and social influence for defying repression) are the deviations from an ideal state for each of the seven different preferences or motivations currently implemented. The weights, $w_x$ for motivation x, give the relative importance of the different motivations to the agent and the relative effect they have on $U$. Different configurations of the weights instantiate different ones of the five social theories described above. These are summarized in Table 5.1. Note that in the ACSES model the weights can assume any value in the range 0 to 1, as long as they sum to 1. In the context of the ACSES model, then, the five theories should be considered preset configurations for the convenience of the user. The user, however, can alter the weights to adjust the theories, to combine them, or even to implement other theories not presented here.

**Table 5.1** Behavioral Theories Implemented in the ACSES Model

| Theory | | Domain | Parameter Settings |
|---|---|---|---|
| Legitimacy | | Obedience | High Loyalty |
| Coercion | | Obedience | High Coercion |
| Representative | | Obedience | High Economics, Violence, Ideology |
| Social Influence | | Social Influence | High Influence + Obedience theory settings |
| Resistance | to | Social Influence | High Repression + Obedience theory settings |
| Repression | | | |

### 5.3.3   The Adaptation Theories

Processes occur at a variety of levels—for example, at the level of the individual agent but also at the aggregate level of society or of a group in society. Moreover, processes at higher, more aggregated levels, while composed of lower level processes, also often affect those processes. The social influence theories explicitly incorporate this phenomenon by having an agent's decision influenced by the aggregated decisions of nearby agents. There is a lower, less aggregated level than the agent that may be important to incorporate as well, especially over the long run. Namely, the situations an agent goes through may affect the agent such that the agent's behavior-choosing criteria change. That is, the agent may adapt. We modeled two of these adaptation processes in the ACSES model as well, using well-known theories of psychological change.

Specifically, the ACSES model incorporates four theories of psychological change for the two behavioral theory parameters that appear most likely to be changeable through the course of a person's experiences: loyalty (the degree to which the person wants to conform to the leadership's orders) and ideology (the side in the conflict that the person's beliefs would lead them to favor). We label the four theories: 1) cognitive dissonance, 2) results-based, 3) homophily, and 4) socialization.

Recall that the loyalty component of the utility function (see equation 2) is *(1-L)*, where *L* is given by the difference between the leadership's orders and the agent's considered behavior, all weighted by the legitimacy of the leadership, η, in the eyes of the agent. In the ACSES model, the cognitive dissonance mechanism (Festinger, 1957) adjusts the perceived legitimacy of the leadership, η, higher if the chosen behavior coincides with the leadership's orders and lower if it does not. The effect of repeated disobedience, therefore, is to lower the perceived legitimacy of the leadership and, thus, lessen the effect of disobedience on the loyalty component in the future.

The results-based mechanism (Coleman, 1990) is based on a measure of the agent's welfare comprising the agent's economic situation and security situation. The perceived legitimacy of the leadership, $\eta$, is adjusted higher if the agent's welfare improves over the previous simulation iteration, lower if the agent's welfare decreases, and remains the same otherwise. Thus, a steady improvement in an agent's situation will produce a large increase in the legitimacy of the leadership and, as a result, increased loyalty by the agent to the leadership's orders concerning allegiance; a steady decline in an agent's situation will have the opposite effect.

The homophily mechanism (Bourdieu, 1984, Carley, 1986) adjusts a citizen agent's perception of the legitimacy of the leadership, $\eta$, to become more similar to how nearby agents like herself perceive the leadership's legitimacy. Specifically, it moves her $\eta$ toward the mean $\eta$ of agents in her ethnic group in his vicinity.

The change brought by the socialization mechanism (St. Augustine, 4th century) is in fact no change at all. It assumes the citizen agent acquired her belief in the leadership's legitimacy in the past, perhaps even when she was a child, and that it is fixed for life. The socialization mechanism is de facto the default psychological change theory in that if no theory is invoked, the perceived legitimacy of the leadership by a given agent will not change over the course of a simulation run.

In the ACSES model, the same four theories of psychological change can be applied to a citizen agent's ideology. Recall that the ideology component of the utility function (see equation 2) is given by $(1 - I)$. $I$ is the discrepancy between the citizen agent's basic ideological position $P$, on a continuum from pro-Taliban through neutral to pro-government, and the choice of allegiance in question. The theories of psychological change concern how P may change over time. The cognitive dissonance mechanism adjusts $P$ in the direction of the agent's chosen behavior. The results-based mechanism adjusts $P$ toward whoever controls the agent's patch if the agent's welfare has improved and away from whoever controls the agent's patch if the agent's welfare has declined. The homophily mechanism adjusts $P$ toward the mean $P$ of like agents in the agent's vicinity. The socialization mechanism assumes that the adult agent's ideology $P$ is fixed and cannot change.

### 5.3.4  The Relationship between Theories

Five behavioral theories are explicitly implemented in the ACSES model. These concern a behavioral choice of where to place allegiance. The three obedience theories are alternatives. The two social influence theories probably should not be used together, but either would be used with an obedience theory. Using a social influence theory by itself makes little sense, in that there has to be some basis for allegiance other than "what everyone else is doing." Thus at the behavioral level there are 9 different combinations: 3 obedience theories X 3 social influence theories (none, social influence, resistance to repression).

Four theories of psychological change are implemented in the ACSES, for each of two inputs to the utility function that instantiates the allegiance theories: the leadership's legitimacy to the citizen agent, which helps determine loyalty, and the citizen agent's ideology. For a given input, the four theories are alternatives; they could be combined, but necessarily the more important one is the less important the others will be. So we will take them here to be strict alternatives. However, the two areas, ideology and legitimacy of the leadership, are completely independent. Thus, one theory could be implemented in one area and a different theory in the other area.

As a result we have a total of 144 different combinations: 9 combinations for the behavioral theories X 4 theories of change in perceived leadership legitimacy X 4 theories of change in ideology. This number is misleading, however, in that, for example, how perceived legitimacy changes will matter only if loyalty has a relatively high weight in the utility function, as under the legitimacy theory. But if either of the other two obedience theories is used, then how perceived legitimacy changes will have little impact.

It is important to note that the relative weights of the seven elements in the utility function can be set to any values between 0 and 1 as long as they sum to 1. Thus two or more of the obedience theories can be combined, by averaging the weights, or entirely new theories involving these preferences or motivations can be implemented. In a sense, then, a continuum of obedience theories concerning allegiance can be implemented. Lastly it is straightforward, if a bit more work, to add a new motivation or preference to the utility function. This may be necessary to implement some alternative behavioral theory not considered for the ACSES model, especially if the target behavior is changed from citizen allegiance to something else.

# Chapter 6
# The ACSES Model of Afghanistan: The Model Operation, Synthetic Population, Calibration, and Surprises

Joseph Whitmeyer

University of North Carolina, Charlotte, NC, USA

## 6.1 Introduction

In this chapter, we continue the description of the ACSES model of citizen allegiance in the presence of an insurgency in Afghanistan. Here, we show how the social theories and actor models presented in the previous chapter were implemented for the specific case of the Taliban insurgency in Afghanistan in an agent-based model. We address four main topics. First, we present and narrate a flowchart of the model. Second, we describe creation of the synthetic population, so that the model would have some fidelity and relevance to the target scenario. Third, we discuss how we calibrated the model using real-world data, concentrating on the genetic algorithm technique that was most fruitful. Finally, we discuss some instances of surprise that we found using the ACSES model, as an example of some of the benefits of this approach.

## 6.2 Geographical Representation

Two-dimensional space is treated in Netlogo as a collection of square patches. We used 865 patches to model Afghanistan. We divided the patches into six regions as shown in Table 6-1, so that we could have six Pashtun leaders, each with influence in its own region and none in the other regions. Note that there are other ethnic groups, much smaller than the Pashtuns. In the simulation, each of these groups had a single leader and the leader has influence on the agents of its ethnic group only, across the entire country.

## 6.3 Simulation Steps

Figure 6.1 presents a conceptual flowchart of the simulation process. We follow it with a description of these simulation steps.

M. Hadžikadić (eds.), *Managing Complexity*,
Studies in Computational Intelligence 504,
DOI: 10.1007/978-3-642-39295-5_6, © Springer-Verlag Berlin Heidelberg 2013

**Table 6.1** The NetLogo Regions

| Region | Name | Number of Patches |
|--------|------|-------------------|
| 1 | Northeast | 155 |
| 2 | Mideast | 195 |
| 3 | Northwest | 93 |
| 4 | Midwest | 92 |
| 5 | Central | 199 |
| 6 | South | 131 |

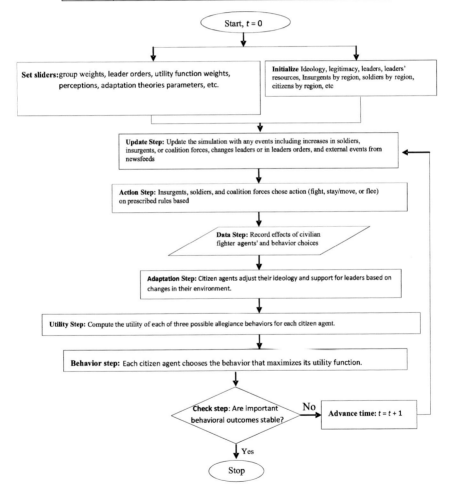

**Fig. 6.1** Conceptual Flow Chart of the Simulation

## 6.3.1   Update Step

In this step, user-initiated, i.e., exogenous changes are made to the simulation. These changes usually will be the insertion or deletion of some kind of fighters into a region, change in a leader's orders, or change in a leader's resources. The change in fighter numbers would represent surges, counter-surges, or withdrawals of forces. The change in leader orders could represent a change of heart by an existing leader; alternatively it could represent the replacement of a leader with certain sympathies with another leader with different sympathies. A change in leader resources could be a consequence of the replacement of a leader, but it could be also that support for the leader has changed, whether from the leader's constituency or from external sources, such as the national government, or foreign powers.

## 6.3.2   Action Step

This step simulates the actions of fighter agents—AGF (Afghan government forces), CF (coalition forces), and Taliban—on each patch. As described in more detail in the previous chapter, for each kind of fighter agent, there are four possible actions: to fight, to make an organized move, to stay on the patch but avoid fighting, or to flee.

## 6.3.3   Data Step

The outcomes of the Action Step are recorded. This includes the casualties, movement of agents, ideology of agents, political allegiance of citizens, and territorial control.

## 6.3.4   Adaptation Step

As described in the previous chapter, among each citizen agent's attributes are ideology—from pro-Taliban (-1) to pro-government (+1)—and loyalty to its group leader (if non-Pashtun) or regional Pashtun leader (if Pashtun). These attributes are given initial values, set by sliders. If an adaptation theory for one or both of these attributes is activated, however, then, in this step, citizen agents adjust their value on that attribute based on the latest developments on their patch, given their previous choices. This step incorporates the adaptation theories described in the previous chapter.

## 6.3.5   Utility Step

At this step, given the updated ideology and loyalty from the learning step together with the economic, coercion, security, repression and social influence variables, each citizen agent computes the relative utility of each of the three behavioral choices he or she has: be neutral (0), help the AGF and CF (+1), or help the Taliban (-1). For the utility function the ACSES model uses the Cobb-Douglas function given in equation (2) of the previous chapter, which we repeat here for convenience

$$U = (1-L)^{w_L} (1-C)^{w_C} (1-I)^{w_I} (1-E)^{w_E} (1-V)^{w_V} (1-F)^{w_F} (1-R)^{w_R}. \quad (1)$$

It may be useful for the reader to see how the utility function is evaluated in more detail (if these details are not of interest, the reader may skip ahead with no loss).

We present each component in turn. Note that each component is given as (1 − X), where X may be considered the diswelfare, possibly experienced as distress, characterizing the agent in that component.

$C$ = diswelfare experienced by the citizen agent due to coercion from the leader. $C$ is given by

$$C = D\,|\,O - B\,|\,/2 \quad (2)$$

where $O$ indicates the leader's order (-1, 0, or 1), $B$ represents the citizen agent's behavior under consideration (recall that the agent will consider -1, 0, and 1 in turn), and $D \in [0,1]$ is the relative magnitude of the resources that can be brought by the leader to bear against an agent. $D$ is set using a slider.

$I$ = diswelfare experienced by the citizen agent due to the difference between the agent's ideology and the course of action the agent is considering. $I$ is given by

$$I = |P - B|/2 \quad (3)$$

where $P \in [-1,+1]$ denotes the ideology of the agent and, as before, $B$ is the considered behavior.

$E$ = diswelfare experienced by the citizen agent due to the agent's economic situation. $E$ is given by

$$E = (E^+ - EB)/(E^+ - E^-) \quad (4)$$

where $E^+$ denotes the best possible feasible economic situation for an agent, $E^-$ denotes the worst possible economic situation for an agent, and $EB$ denotes the economic situation the agent expects by following the given considered behavior. If the citizen agent belongs to an area where opium is grown (a 38% chance on "opium" patches), then $E^+ = \$2000$; if the citizen agent belongs to an area where opium is not grown, then $E^+ = \$300$; and the worst economic situation all agents expect is $E^- = \$50$.

$V$ = diswelfare experienced by the citizen agent due to the agent's security situation. $V$ is given by

$$V = \frac{v_c}{v_C^-} \qquad (5)$$

where

$$v_c = \frac{foes}{friends + foes} \qquad (6)$$

and $v_c^-$ is the worst possible value of $v_c$.

$R$ = diswelfare experienced by the citizen agent due to repression by armed forces (Taliban or pro-government forces) at the patch level.

Repression as a component is activated only when a citizen agent is considering helping the side that is outnumbered on the agent's patch. That is, if the considered behavior is neutrality or helping the predominant force on the agent's patch, $R$ is set to 0. Otherwise, the defining characteristic of repression is that it is less effective the more that other citizen agents in some circle around the deciding agent are defying repression. Thus, a citizen agent is less likely to defy a faction that has a relatively strong presence on the agent's patch, especially when that faction has little support among other citizen agents in the general area. $R$ is computed as follows:

$$\kappa = \frac{n_S + n_C - n_T}{n_S + n_C + n_T + n_I} \qquad (7)$$

where

$n_S$ = the number of government soldier agents in the focal agent's patch,

$n_C$ = the number of coalition soldier agents in the focal agent's patch,

$n_T$ = the number of Taliban insurgent agents in the patch, and

$n_I$ = the total number of citizen agents in the patch.

Positive values of $\kappa$ indicate that the AGF and CF outnumber the Taliban, whereas negative values indicate the opposite. The magnitude of $\kappa$ indicates the degree to which one faction outnumbers the other. $R$ is then given by

$$R = \max[0, -Sign(\kappa) \times Sign(B)] * \max[\kappa^2 - \frac{1}{2}\left(\frac{1}{n_I}\sum_{i \in n_I} B_i\right)^2_{t-1}, 0] \quad , \qquad (8)$$

where $\left(\frac{1}{n_I}\sum_{i \in n_I} B_i\right)_{t-1}$ = the proportion of citizen agents in a larger geographical area (one or more patches) performing behavior $B$ in the previous time step (i.e., $t-1$).

The first factor in equation (8) is 0 unless the considered behavior opposes the dominant force on the patch. The second factor increases with the predominance of the fighters of one side and decreases the more citizens in the given area are performing the behavior this particular agent is considering.

For example, suppose that for a given patch, $n_S = 0$, $n_C = 0$, $n_T = 200$, $n_I = 50$, and all citizens were neutral in the preceding time step. Then $\kappa = -200/250 = -0.8$. For behavior $B = +1$ (support soldiers)

$$R_{(B=+1)} = \max[0, -Sign(-0.80) \times Sign(+1)] * \max[(-0.80)^2 - \frac{1}{2}(\frac{0}{50})^2, 0]$$

$$= \max[0,1] * \max[0.64, 0] = 0.64$$

Suppose 30 citizens support the Taliban. Then for behavior $B = -1$ (support Taliban)

$$R_{(B=-1)} = \max[0, -Sign(-0.80) \times Sign(-1)] * \max[(-0.80)^2 - \frac{1}{2}\left(\frac{-30}{50}\right)^2, 0]$$

$$= \max[0, -1] * \max[0.46, 0] = 0$$

For the above example, if $w_R = 0.2$, for support soldiers $(1-R)^{w_R} = (1-0.64)^{0.2} = 0.815$, whereas for support Taliban, $(1-R)^{w_R} = (1-0)^{0.2} = 1$, which implies that for this patch, repression hinders support for soldiers but does not hinder support for the Taliban.

$F$ = the diswelfare experienced by the citizen agent due to the social influence of close associates.

We can consider this diswelfare to be a function $F(B)$ of the considered behavior $B$. Specifically, then, $F(B)$ = the deviation of the citizen agent's behavior from the behavior of citizen agents within a certain radius of the focal agent in the previous time step (i.e., $t-1$). $F(B)$ is computed as follows:

$$F(B) = \left| B - \left(\frac{1}{n_I}\sum_{i \in I} B_i\right) \right|, \tag{9}$$

where $I$ refers to the agent's patch, $i$ is a fellow citizen agent on the patch, and $B_i$ refers to fellow agent $i$'s behavior the previous time step.

For example, if there are $n_I = 400$ on a patch and 100 of them supported the soldiers last time and 20 supported the Taliban in the last time step, then for "support Taliban" behavior

$$F_{B=-1} = \left| -1 - \frac{-20}{400} \right| = 0.95, \quad \text{and} \quad \text{for} \quad \text{"support the soldiers"}$$

$$F_{B=+1} = \left| 1 - \frac{100}{400} \right| = 0.75 .$$

If $w_F = 0.2$, then for "support soldiers" $(1-F)^{w_F} = (1-0.75)^{0.2} = 0.758$, whereas for "support Taliban," $(1-F)^{w_F} = (1-0.95)^{0.2} = 0.549$, which implies that for this patch, influence works to increase support for soldiers relative to Taliban.

## 6.3.6   Behavior Step

In this step, each citizen agent picks the behavior that maximizes her utility, as computed in the Utility step. Every Taliban helper and soldier helper has a force weight that indicates his or her relative contribution to the battles. Therefore, this aggregate behavior of the citizen agents influences the subsequent action choices for the CF, AGF, and Taliban by affecting the perceived effectiveness for each. It also contributes to the real effectiveness of each, which is used to determine the Action step's outcomes.

## 6.3.7   Check Step

In this step, three simulation outputs—territorial control, hardship, and political allegiance—from the current step are compared to those from the previous steps to determine if all of the outcomes have become sufficiently stable to terminate the simulation. The length of the simulation run will depend on the frequency and the significance of the events and changes inputted in the Update step. The simulation outputs are calculated as follows.

Territorial control refers to control at the patch level:

$$\text{Control} = \frac{n_S \times e_S + n_C \times e_C - n_T \times e_T}{n_S \times e_S + n_C \times e_C + n_T \times e_T} \tag{10}$$

where $n_S \times w_S + n_C \times w_C$ is the effectiveness of soldiers and $n_T \times w_T$ that of the Taliban.

For hardship:

$$\text{hardship} = \text{clip-to-range}_0^1 \left( \frac{\text{security} \cdot V + \text{economics} \cdot E}{\text{security} + \text{economics}} \right) \tag{11}$$

Here, "security" and "economics" refer to the relative importance of security and economics and are set by sliders. Clip-to-range is a "squashing" function such as for the instance of the sigmoid function. This is done to keep the values in range 0 and 1 as shown in the figure below.

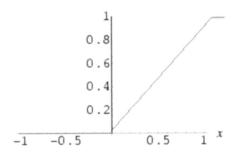

**Fig. 6.2** Clip to range function

Political allegiance refers to which side the citizen agents support on the aggregate. It is calculated on the patch level, based on how many of the citizen agents choose to support the government, stay neutral, or support the Taliban. The aggregate of this attribute is used to track and measure certain regional effects. Political allegiance that is heavily tilted one way or the other can also influence the outcome of combat. Political allegiance is calculated as follows:

$$\text{Political allegiance} = \frac{n_{SH} - n_{TH}}{n_N + n_{SH} + n_{TH}} = \frac{n_{SH} - n_{TH}}{n_I}. \tag{12}$$

Note that if the purpose of a simulation run is training and testing the relationship between simulation outputs and real-world events, then we are constrained to using the putative relationship between the simulated and real-world clock. In this case the stabilization criterion is no longer used because these experiments constrain the total number of simulated time steps that are allowed to elapse to represent the real time that transpired.

## 6.4    The Synthetic Population

The foundation of the ACSES simulation model is a synthetic population with attributes important to the modeling objectives. We constructed a synthetic population representative of Afghanistan's 2006 population. We describe, here, the data sets used   and how they were disaggregated and integrated to populate the model with agents having the desired attributes.

### 6.4.1    Data Sets

The social theories implemented in the ACSES model require input on loyalty, coercion, ideology, economic welfare, and security from violence for each agent of the population, i.e., citizen agent. To geospatially assign attributes for the agent's support for the leadership, the leadership's resources, the agent's ideology, and the agent's degree of satisfaction with the economic and security situations, databases and data sources were employed according to the following flow diagram.

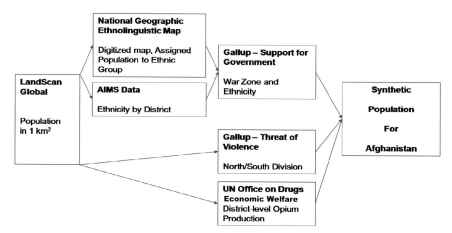

**Fig. 6.3** The flow of data sources employed in construction of the Afghan synthetic population

The first step in building the synthetic population was geospatially locating the 31 million agents representing the Afghan population within grids approximately one kilometer on each side. LandScan Global techniques for disaggregating a population produced the geographical population distribution illustrated below.

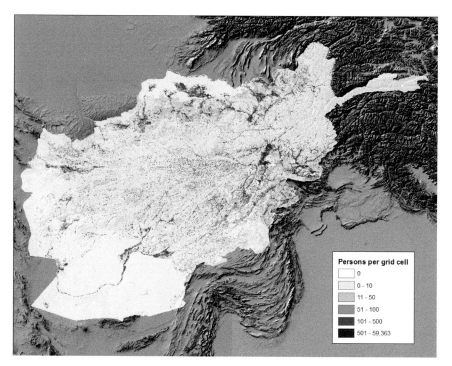

**Fig. 6.4** Geo-spatial distribution of agents representing the Afghan population

Assigning attributes to this disaggregated population was achieved by fusing appropriate indicator databases. A diverse body of literature describes the major Afghan political and social cleavages that can be used to deduce ideology and perceptions of security against violence. We began by consulting district profiles collected through human-terrain-type efforts by the Afghan Information Management Service (AIMS) from which we can determine the ethnicity of each element in 206 of the 329 Afghan districts. For the remaining 123 districts, the ethnic breakdown was estimated by interpolating data on ethno-linguistic maps, such as the one below, that delineate the distribution of ethnic groups. Ethnic data from adjacent districts was used for the interpolation when it was available. Otherwise, data from nearby areas sufficed. Although many granular maps exist, the National Geographic maps were used to demonstrate the value of common maps of limited granularity.

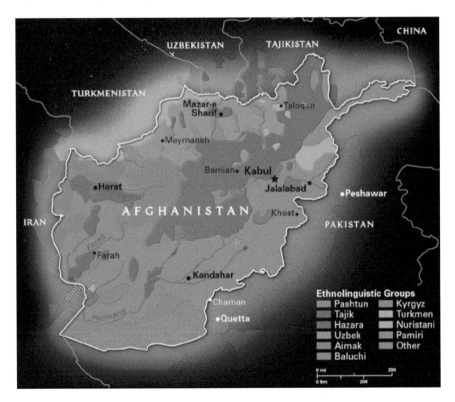

**Fig. 6.5** Ethno-linguistic mappings used for ethnic interpolation of ethnicities in missing districts. Source: National Geographic.

Once the ethnicity for each of the 31 million agents was assigned, then the ideology and attitudes toward the economic and security situation were assigned to each social atom. Although very specific polling data can be gathered in areas of high U.S. national interest, we used existing Gallup World Poll data for this study to illustrate the methodology.

The attitudes within the Gallup study group seem to have significant cleavages along ethnic (Pashtun, Tajik, and Other), location (north, south), and combat proximity (war zone vs. non-war zone) lines. A mapping for war zone location is reproduced below.

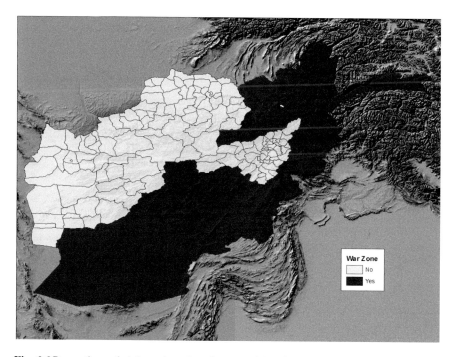

**Fig. 6.6** Perceptions of violence based on the agents' location respective of the war zone

The indicator database for how economic welfare is perceived by the individual and is tied to behaviors in support of leaders in the Afghan case is highly dependent upon the social atom's proximity to opium production. The Afghan example is simpler than most countries where corruption indices might create usable indicator databases and where economic activity under leadership control may be more complex. A map of opium districts is shown below.

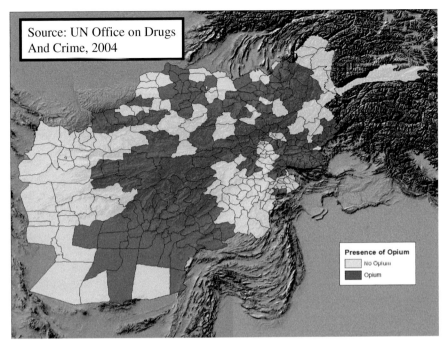

Source: UN Office on Drugs
And Crime, 2004

Presence of Opium
No Opium
Opium

**Fig. 6.7** The areas of opium cultivation from UN Office on Drugs and Crime data from 2004

The elements above, taken together, were the building blocks of our synthetic population.

## 6.5   Calibration of the ACSES Model

These specific simulations are based on the situation from 2005 to 2007 in the Ghazni region of Afghanistan, about halfway between Kabul and Kandahar. The region was chosen because of a high level of Taliban activity in that period and a counterinsurgency operation (Operation Maiwand) by the Afghan government and coalition forces in June 2007. The Taliban use removal and intimidation of the local leadership as a tactic to increase their control over the population. As one of our input variables, therefore, we use assassinations or attempted assassinations of local leaders. Our output variable will be the allegiance of the citizen population.

The simulation is trained on data covering seven months in 2007 (3/15/2007 to 10/15/2007, to avoid a seasonal effect in violence unaccounted for by our simulation). The year2007 was selected because it contains a good variety of input data events on which to train the simulation.   For these data we used the number of terrorist incidents in Afghanistan as reported in the Worldwide Incidents Tracking System, or WITS, database.

The three significant components of the calibration process, in which key parameter values are tried and evaluated, are the inputs, the outputs, and the correlation measure between the simulation and the WITS data.

## 6.5.1  Inputs

The simulation accepts both static inputs and variable inputs. The static inputs are those that come directly from data, including the initial population settings as well as events that were known to transpire during the simulation period. The variable inputs are controls that we perturb during calibration to identify useful settings.

### 6.5.1.1  Static Inputs

Most of the static inputs are elaborate elsewhere in the chapter. Briefly, they include

- Geography
- Population per patch (including religious affiliations and tribal memberships)
- Hand-selected political events (assassinations of prominent figures; see below); not all political events have direct analogues within the current simulation, but assassinations do, and all were incorporated into the event stream
- Perceptive acuity for both factions: this is the likelihood of each faction correctly identifying its own units and its opponents' units, expressed in a matrix of sliders within the NetLogo simulation
- Effectiveness measures for both factions
- Leaders' resources and standing orders (coarse allegiance)
- Periodic infusion rates of new soldiers and Taliban

### 6.5.1.2  Specific Assassination Inputs

These are the specific assassination events that are introduced into the simulation:[1]

- April 16th, 2007 – In Andar, Ghazni, Afghanistan, assailants launched rockets near a high school, causing no injuries or damage. The attack targeted the governor of the province, who was attending a road construction inauguration event. No group claimed responsibility, although it was widely believed the Taliban was responsible.
- June 1st, 2007 – In Khwaja Umari, Ghazni, Afghanistan, armed assailants abducted the Chief of Gomal District from his house. The Taliban claimed responsibility.
- June 2nd, 2007 – In Angor Village, Jaghuri, Ghazni, Afghanistan, armed assailants stormed the house of a police chief then fired upon and killed five civilians. The Taliban claimed responsibility.

---

[1] These descriptions are copied verbatim from the WITS database.

- August 3$^{rd}$, 2007 – Near Ghazni, Ghazni, Afghanistan, assailants abducted a provincial government official. The Taliban claimed responsibility.
- September 3$^{rd}$, 2007 – Near the Leewani Bazaar, in Qareh Bagh, Ghazni, Afghanistan, assailants kidnapped two civilian relatives of a female legislator. The Taliban claimed responsibility.

The simulation assumes that Taliban assassination of pro-government leaders will have the effect of moving the aggregate leadership toward a pro-Taliban stance. Each act of aggression by the Taliban against that leader will move the leader's order toward -1 and away from +1.

Furthermore, the leader's order is adjusted according to which side (the Taliban or the government) controls the Ghazni patches. Control is measured using the number of, and the force weights for, coalition forces (CF), Afghan government forces (AGF), and Taliban, such that

$$\text{Soldier-Force} = SF = e_C \times n_C + e_S \times n_S + e_{SH} \times n_{SH}$$

$$\text{Taliban-Force} = TF = e_T \times n_T + e_{TH} \times n_{TH}$$

$$\text{Control} = \frac{n_s \times e_s + n_c \times e_c - n_T \times e_T}{n_s \times e_s + n_c \times e_c + n_T \times e_T}$$

where the $n$ values are the number of different factions present, the $e$ values are the effectiveness of each faction, and the S, C, and T subscripts refer to the factions themselves.

The leader's order is adjusted by control using the following function:

$$O_{t+1} = O_t + (1 - O_t) \times \max(0, Control_t)$$

To understand this equation and why Taliban control does not influence the order, think of the status quo ante as the baseline: the leadership's order with no assassinations by the Taliban and no government troop surge. Then the Taliban and the government each have an activity available to them to affect the order: the Taliban can assassinate and the government can initiate a troop surge. If the surge fails militarily, meaning that the government fails to achieve control, then the leadership's order should remain unaffected, that is, return to the pre-surge status quo. (The rationale that the Taliban somehow seem more like "winners" by that failure so the leadership should shift to them but because a failed surge could just as easily make a larger surge likely, the leadership would not want to be become more pro-Taliban.)

In other words, control in the area by the coalition forces can affect a leader's orders, but control by the Taliban does not. These equations are incorporated into a flow diagram as follows:

**Table 6.2** Simulation Procedures within the conceptual flow diagram

| Simulation Procedures | Conceptual Flow Diagram |
|---|---|
| **1)** Regenerate new agents – **add surge** | Start, $t = 0$ |
| Time decay of WITS violence per patch | • Set sliders |
| **1)** Input new WITS data – **input assassination data; update leader's order** | • Initialize |
| | • **1) Update step** |
| Small chance of civilians moving to neighboring patch | • Action step |
| | • **2) Data step** |
| Record fighter-agent deaths (previous turn deaths: "violence-last-turn") | • Learning step |
| | • **3) Utility step** |
| Calculate influence and repression for utility function | • **4) Behavior step** |
| | • Advance time or stop |
| Calculate real and perceived effectiveness | $t = t+1$ |
| Commence battle, or troops may move if no battle | |
| Record fighter-agent deaths (deaths this turn: "violence-this-turn") | |
| **2)** Store control, political allegiance, and $E^+$ for each patch – **store control for Ghazni region; adjust leader's order based on control** | |
| Calculate partial security threat for adaptation theory | |
| Run adaptation theory module, which affects agent's support for leaders and ideology | |
| **3)** Calculate remaining factors for utility function – **use adjusted leader's order from above** | |
| **4) Run utility function** | |
| Refresh and visualize effects | |

## 6.5.2   Time and Scheduling

The time scale is one of the parameters we allowed to vary over the simulation runs. We hoped to identify a setting that best contributes to the realism of the simulation output. We can make no claim that the resulting time scale is psychologically or sociologically motivated, only that it was useful.

Later, we describe stabilization criteria that are used to terminate runs. These criteria are appropriate in the context described later, but here we are training and testing the relationship between simulation outputs and real-world events, so we are constrained to using the putative relationship between the simulated and real-world clock. Hence, these experiments constrain the total number of simulated time steps (ticks) that are allowed to elapse to represent the seven months that transpired in the real-world.

The political events are, therefore, scheduled to correspond to ticks on the simulation clock. The following table shows the exact time steps used for events in the simulation runs. This is just an example; clearly, the experiment includes more than seven events in the simulations.

**Table 6.3** Model Event Time Steps

|          | Event #1 | #2 | #3 | Surge | #4 | #5 | End |
|----------|----------|----|----|-------|----|----|-----|
| 3 days   | 11       | 26 | 27 | 37    | 47 | 58 | 68  |
| 7 days   | 5        | 12 | 12 | 16    | 21 | 25 | 35  |
| 14 days  | 3        | 6  | 6  | 8     | 11 | 13 | 23  |

## 6.6   Outputs

The critical output is political allegiance.

## 6.7   Procedures and Results

We pursued three options for describing the agents in our simulation, with the overriding goal of finding a description that produces simulation results most predictive of the real Afghanistan. These options were to use three different procedures for generating citizen agents: one with the social theories, i.e. agent models, set *a priori*; one with the social theories set *a priori* but that allows change via the adaptation theories; and one that consists of free exploration. Not surprisingly, we found that the third produced the best results, so it is that one which we describe here.

The third free exploration approach began with random attribute values and was calibrated by a distributed genetic algorithm (GA) to favor those values in simulations that more faithfully represented the real world. Specifically, we allowed a free search of the space by a distributed genetic algorithm (GA). We now review the technical design, the results and the analysis.

## 6.8   Technical Design of the GA

The core technical challenge was that the NetLogo platform on which the simulation is based is decidedly single-threaded, which stymies parallelism. To work around this, we constructed our GA system so that each processing node ran exactly one instance of the NetLogo simulation. A master node was always responsible for managing the central distribution of work and collecting results from the processing nodes.

The specifics of the GA are as follows:

1. The population size was dictated by the number of processing nodes available, typically between 16 and 24 chromosomes (at one chromosome per processing node).
2. Generations were synchronized by the master node.
3. Each chromosome was a vector of social theory (allegiance) parameter values. Upon evaluation, each chromosome was instantiated in a headless instance of NetLogo, the simulation was run for the requisite 64 time steps, and the correlation between political allegiance and WITS incidents was computed and returned—normalized to (1+CORR)/2—as the fitness value.
4. When computing the next generation, we used single-point crossover (80%) and simple Gaussian mutation (20%).
5. The GA was stopped after completion of 100 generations; the members of the final generation were reported as the GA results, along with their fitness values.
6. A single run required almost 13 hours of clock time using 24 parallel computing nodes.

## 6.9   Results from the GA

After a number of trials, we found that individuals like the one reported below are representative of the best-performing chromosomes discovered by the GA:

**Table 6.4** Best Performing Chromosomes

| Loyalty weight | 0.346 |
|---|---|
| Coercion weight | 0.249 |
| Ideology weight | < 0.001 |
| Economics weight | 0.096 |
| Security weight | 0.000 |
| Influence weight | 0.167 |
| Repression weight | 0.141 |
| Correlation | 0.152 |
| Fitness (1+CORR)/2 | 0.576 |

The course of the simulation is expressed in the following two charts taken directly from the NetLogo display.

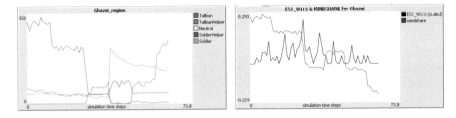

**Fig. 6.8** Course of Simulations

The left chart displays the composition of the Ghazni population over time. Most notable in this chart is the surge of soldiers corresponding to operation Maiwand. The right chart displays the two series we are trying to correlate: political allegiance ("mindshare") and the number of WITS incidents within Ghazni.

The most significant finding is that the weights for both ideology and security are essentially zero. This is consistent with our findings in Approach 1 (not presented here), in which the legitimacy theory of allegiance was identified as the best (best average) performer; the legitimacy theory also specifies that the security weight should be zero.

It is interesting to note the difference between political allegiance and control in the simulation for this winning chromosome. Consider these two images directly from the NetLogo simulation:

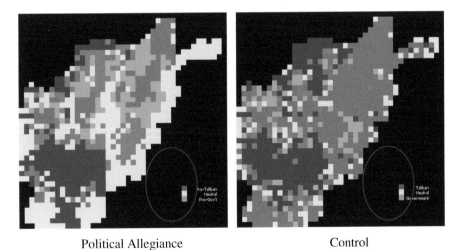

Political Allegiance                                    Control

**Fig. 6.9** The difference between political allegiance and control in the simulation

Ghazni, loosely represented by the red oval superimposed on the two images, has been strongly influenced toward a pro-Taliban allegiance, yet it remains firmly in control of the coalition and pro-Afghan government.

We also conducted, but do not report here, additional work in which we ran the GA assuming that simulated political allegiance and the number of WITS incidents were anti-correlated. Expectedly, a different set of parameter values bubbled to the top of the list of most fit chromosomes. Of greater interest, though, was the fact that those runs also highlighted effects on political allegiance resulting from the Maiwand surge that were not visible when the correlation was presumed to be positive: When the troops surged, the populace responded by rallying in a pro-government manner.

## 6.10   Analysis of the GA

The GA demonstrated the following advantages over both Approach 1 (static, predefined test cases) and Approach 2 (static cases with adaptation):

1. The combinations of parameter values that we explored were significantly more varied than those we had hand-selected for Approach 1. Flexibility in choosing parameter combinations is especially important in a search algorithm operating in a search space as significantly multimodal as this simulated Afghanistan. Adaptation succeeds best only when it can overcome the exceedingly difficult problem of avoiding local minima, but without knowing the exact bounds on solution quality and volatility, a grid-based search is not appropriate.

   It is not reasonable to claim that the GA does any better job of recreating ground truth as observed in the real world Afghanistan than does the set of discrete, static tests employed in Approach 1. Instead, the claim is merely that – not knowing precisely how well these simulations reproduce what happened in the real world – we find that the advantage within the GA implementation is that it is free of the arbitrary constraints that underlie the selection of discrete simulations tested in Approach 1. The GA, then, is subject to far fewer social theory assumptions than is either of the preceding approaches. While it is certainly possible that this flexibility is not required for this specific simulation, especially if there is some innate resilience within the simulation to change in light of varied parameter settings, the generality of the approach has merit.

2. Even though the parameter values were unconstrained, as mentioned in the previous point, the results achieved were consistent with the conclusions drawn in Approaches 1 and 2: the simulation performs most successfully when security weight is low.

3. Although the number of potential candidates (chromosomes) to evaluate is much higher with a GA than with either of the other two approaches described above, the confidence we have in the corroborating results

among the three methods makes this increased computational effort worthwhile.

A few additional insights from this analysis were as follows. First, it is possible, even within a simulation that relies heavily on stochastic processes, to evaluate multiple replications of discrete combinations of parameter values to identify optimal simulation settings and trends. Second, the simulation can exploit the underlying social models. This is a direct consequence of the equations embodying the social models, rather than an emergent behavior of the society of agents. Third, adaptation can increase the verisimilitude of the running model in some cases.

## 6.11   Accounting for Surprise

Often the purpose of an agent-based model, such as the ACSES model, is twofold: to explain the causal reasons for known significant events and to predict unknown significant events. In the first case, we must know the environment and the events in order to hypothesize the causes; in the second case, we create a likely or interesting environment, and then we use the hypothesized causes to see what will happen here. Many of the underlying causes may be variables that cannot be readily measured. So we create agents that have certain properties—both attributes and rules—and then we calibrate the attribute values and the rules so that the outputs of the system match the macro-level events that we can measure. These refined attribute values and rules become our hypothesis to be used in the prediction stage.

The surprise is also twofold: if we are successful, we have support for inferring the underlying causes of the events, which we did not know previously. Furthermore, given these underlying causes, we can find out what will happen in a different environment, either under new conditions or during simulated intervention. The key element in "surprise" is that we have measurable, macro-level events that are not explicitly programmed. Surprise is therefore emergent from the interaction of attributes, values, and rules at the agent level.

Below we describe two examples of surprise obtained with the ACSES model, the macro-level agent behavior of clustering, and a threshold effect when varying certain variables. Neither of these effects is explicitly programmed into the model.

Clustering is representative of *social behavior emergence*; that is, it is a self-organizing phenomenon that is a consequence of the agents' individual actions and changes in state as they react to other agents and to the environment. Recall that citizen agents and fighter agents differ in important ways in the ACSES model. Citizen agents can change allegiance but cannot move. Fighter agents belong to one of three groups; the Taliban, Afghan government forces (AGF), or coalition forces (CF); and cannot change affiliation. Fighter agents can choose at each time step whether to fight, move, stay (on the current patch), or flee, making their choice based on their perception of their own effectiveness relative to the local *perceived* effectiveness of the opposing force. These simple rules produce effects in the system that are not explicitly accounted for, but that occur as a natural consequence of these rules. One such effect is that both the soldiers and the Taliban will begin to group together. The fighter agents are more likely to survive in large groups. Hence, larger groups will emerge in the system. No rule explicitly tells the fighter agents to group together; it just happens.

Below are images and statistics from a single, but typical, simulation run. The brown dots represent Taliban agents, the green dots are soldier agents, and the other colors represent civilians—blue for civilians who are pro-government, yellow for those who remain neutral, and red for those who are pro-Taliban. The first image (Fig. 6.10) is from the initial setup for the run. The civilians are placed according to demographic information on Afghanistan, while the soldiers and Taliban (1500 agents each) are allocated across Afghanistan randomly. Notice that the placement or grouping of the green or brown agents follows no discernible pattern.

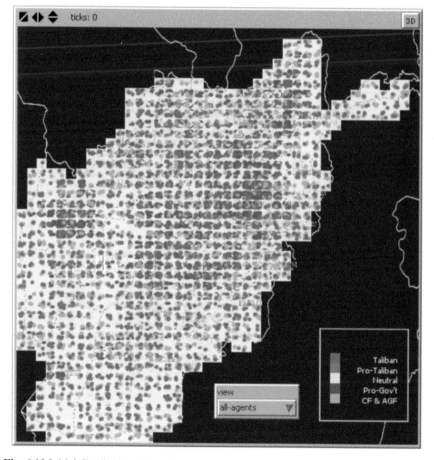

**Fig. 6.10** Initial distribution of agents

After 50 time steps in the simulation (Figure 6.11), some patterns begin to emerge. Of course, many of the Taliban and, to a lesser extent, the soldiers have disappeared from the simulation. This die-off of agents occurs in the early stages of the simulation run as, presumably, many of the weak and isolated agents have been eliminated. The Taliban do not have the same force strength or success rates as the CF agents. Thus, their elimination rate is more pronounced. Note that in these simulation runs we inject additional fighter agents randomly into the system to preserve some stability in the number of these agents.

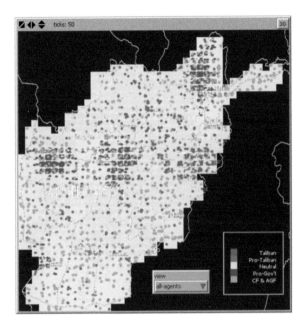

**Fig. 6.11** Distribution of agents after 50 time steps

By time steps 150-200, stronghold areas for the Taliban always emerged. Figures 6.12 and 6.13 show two different runs. Around time step 350, the eastern central area is always one area of Taliban strength. At times it is the only one; some runs also have the northwest region; others have the southern region, as seen in Figure 6.12. Clearly, the outcomes of the simulation runs are not always the same. Nevertheless, they illustrate the fact that the Taliban agents tend to cluster together, a trait that is not explicitly programmed into the simulation. Again, our hypothesis is that this is because this increases their chance of survival.

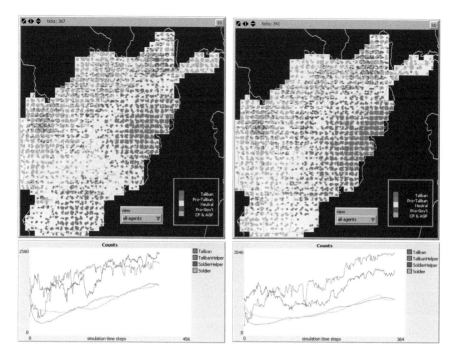

**Fig. 6.12** Stable results from a run          **Fig. 6.13** Stable results from a run

In contrast to phenomena like clustering, *threshold* effects only become apparent by comparing multiple runs of the simulation, varying one or more parameters slowly. Here, we show the effects on the number of soldier helpers of increasing the size of a surge in coalition forces into the Ghazni region (in Eastern Central Afghanistan). This case was chosen by taking one set of settings from the design used in the calibration. The specific settings were: Time Scale = 3; Social Theory = Representative; Influence = On; Repression = Off. At startup, the numbers of CF, AGF, and Taliban were 1000. No agents were injected at each time step, so the region had only its share of the initial soldier forces supplemented by the subsequent surge of coalition forces.

We started with a surge level of 40 and increased that by 10, up to 110. Each of these surge levels was run 60 times. We also ran three different mid-range surge levels per interval, 30 times each (i.e., we ran surge levels of 42, 45, 48 and 52, 55, 58, etc., up to 102, 105, and 108). We defined a surge as successful if the number of soldier helpers at the end of the run was greater than the number of soldiers. The following graph is a plot of the number of runs that met the criterion for a successful surge. In this graph, the count results from runs of 30 repetitions are doubled to normalize with results from runs of 60 repetitions.

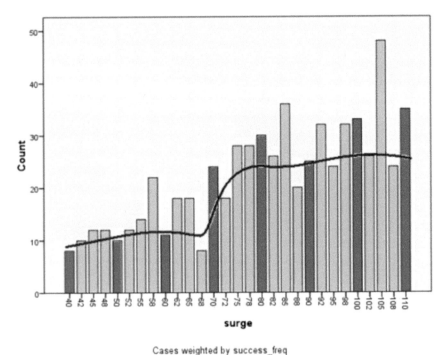

Cases weighted by success_freq

**Fig. 6.14** Threshold effects

There is considerable variation in results for a given set of parameter settings. Nevertheless, as the red line in Figure 6.14 makes apparent, when the surge level reaches approximately 70 soldiers, the number of successes increases dramatically. With a few exceptions, when the surge is less than 70, the number of successes tends to be around 10 to 15 out of every 60 runs; when the surge level is greater than 70, then the success rate tends to be 25 to 30 out of every 60 runs. Some sort of threshold is apparently reached at the 70-soldier surge level that allows the system to greatly increase the success rate. An undoubtedly related effect of reaching the 70-soldier level is that a strong separation between success and failure emerges at that surge level: the number of soldier helpers is low or high, not medium. This is suggested in the histogram in Figure 6.15.

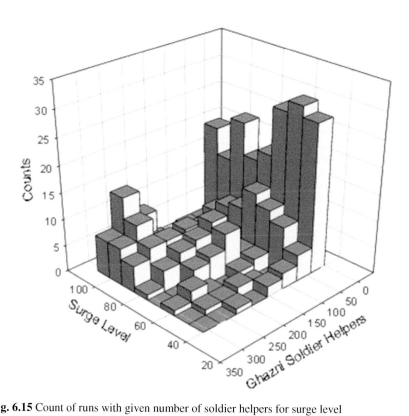

**Fig. 6.15** Count of runs with given number of soldier helpers for surge level

# Chapter 7
# Implementation Influences

Mark Armstrong

IntePoint Corporation, Charlotte, NC, USA

## 7.1 Overview

The usage of agent based models for analysis is transitioning from research laboratories to applied capabilities. As the audience of end users changes, the choices of implementation decisions have significantly more impact on the usability of the analysis.

This section provides a brief overview of some of the key decision points and their importance when designing an ABM for deployment and use by personnel other than the simulation developers.

## 7.2 Audience

It is critical to have a good understanding of the ways that the ultimate end users will leverage the agent-based model as part of their tasking or decision-making.

Historically, these models have been used as a component of research activities. In this role, it is acceptable if the results are only demonstrable, analysis of results is performed manually by the researcher(s), changes to the scope of the model are suggested by the researcher(s), the output is captured and described for a publication, and the time scale of the project ranges from months to years.

Over the last 5+ years, multiple non-research communities have begun to embrace the value of agent-based models as a key element of their production workflows. Shipping cargo fleet scheduling by Tankers International (Himoff, 2005), pipeline optimization, aerospace contract management, and recent progress in economic modeling provide a few examples of increased adoption of agent-based modeling. The Department of Defense is also expanding its use of ABM for the purposes of social/population behavior impact modeling.

If the end state of the analysis is for personal use or to support research, then the design constraints are considerably looser. Essentially, the creator of the simulation has ultimate freedom in addressing implementation design elements.

In the scenario where the results of the simulation are to be used by others as a supporting element of their decision making, efforts must be made by the

M. Hadžikadić (eds.), *Managing Complexity*,
Studies in Computational Intelligence 504,
DOI: 10.1007/978-3-642-39295-5_7, © Springer-Verlag Berlin Heidelberg 2013

simulation designer to have the modeling, simulation and analysis features fit seamlessly into the user's process and time scale.

## 7.3   Design Influencers

A full discussion about the approaches to delivering a useful capability is well outside the scope of this book. The following comments address specific design issues that are relevant to ABM.

Designers are well advised to follow the advice from Stephen Covey (2004) to "begin with the end in mind." An ABM can be a complicated system with many options for adjusting parameters, visualization choices, validation and verification strategies, etc. The user interface can easily become overloaded with displays, buttons, and sliders. Often, a single interface is used for model building, simulation execution, and output analysis, thus cluttering the user interface with many unnecessary artifacts.

In order to avoid some of these pitfalls, it is necessary to have an understanding of how the end user(s) will interact with the system. Some key design questions follow:

1) Who will be the primary users and consumers of the simulation and its outputs?

   In organizations that leverage modeling and simulation as a component of their workflow, there are typically four roles that influence the use of a specific model, data collection, model building, simulation analysis, and decision-making. Each role has distinct requirements and may benefit from different user interfaces to the agent-based model.

   a) *Data Collection.* The person in this role must understand the model in enough detail to identify and procure the data required to run the simulation. In addition to the data source input interfaces, this person may benefit from sensitivity and relationship diagrams that help him make choices on data quality and quantity. The end user organization will be required to have a data collection process.

   b) *Model Building.* The model builder is responsible for creating the functions that support the simulation runs, see Figure 7.1 for example. This person is analogous to that of a software developer combined with a software architect and is responsible for the design and implementation of technical underpinnings of the simulation. The model builder typically works through the technical interface of the agent-based model. There is not a requirement for the end user organization to have model builders on staff.

**Fig. 7.1** An example of programming functions in NetLogo

c) *Simulation Analysis.* In this role, the analyst/operator is required to both set parameters and translate the simulation results into actionable knowledge. To accurately use the same model for different problems, e.g. traffic flow prediction in Los Angeles instead of Dallas, the parameters of the simulation must be adjusted. In a complex simulation, there may be hundreds of parameters. For this aspect of the role, the analyst/operator would benefit from a user interface that clearly defines each parameter and provides feedback to the analyst/operator about the impacts of any changes to those parameters.

Simulation results analysis requires a different set of features on the user interface. Rather than being shown all of the parameters, it may benefit the analyst/operator to have tools that allow them to explore and explain the results, e.g. drill down data, threshold alerts, multiple graphs (2D, 3D, network, multiple formats). Figure 7.2 provides an example of such an interface developed for the ACSES project. The focus of this element of the workflow is to extract actionable knowledge from the simulation that can be presented to the decision makers.

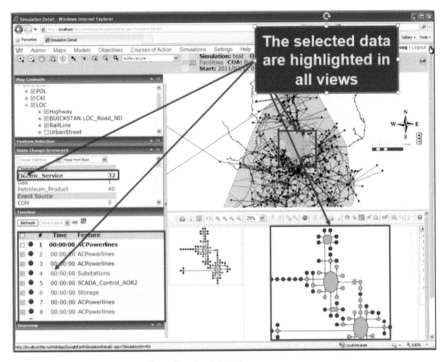

**Fig. 7.2** An example of a user interface in ACSES

    d)  *Decision Making.*  Data, internal model relationships, parameter settings, and raw simulation output are interesting but irrelevant for decision makers. They want concise, preferably visual, output from the model that can be seamlessly incorporated with the rest of the data points from other sources, so that they and their team can make actionable decisions from that information. The effective, meaningful visualization of the results of agent-based simulations is difficult to achieve and often hampers the acceptance of the information provided by the model (see Figure 7.3).

        Design of the mechanism for the dissemination of the results is essential to the long term acceptance and usage of agent based models in non-research scenarios. It is also often appended at the end of the project by the model team having lost sight of Covey's axiom.

2)  What is the length of the analysis cycle that makes use of the agent based model?

Time-to-results will be a key performance metric that impacts the end user organization's perception of the value of the agent based model as a key analytics capability or a tool that generates interesting results but is not relied upon for decision making. Regardless of the effectiveness of the user interface design, ease of use, and presentation of results, if the decision making cycle for the end user is 2 days and it takes 2 weeks to run a simulation, the results will never be incorporated as a mission critical capability.

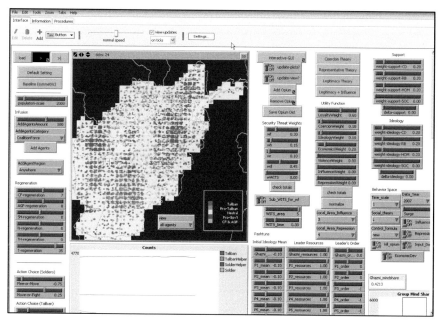

**Fig. 7.3** An example of a "developer's" interface

There are several areas where the internal design of the agent based model affect performance:

a) *Computing power.* This question boils down to how much hardware can the end user allocate to the running of a simulation. Large financial institution or government agencies are able to provide a high-performance computing (HPC) platform. In this scenario, the choice of agent platforms becomes a key element of the design decisions, as not all agent platforms can leverage HPC environments. Additionally, the model building team will require HPC programming talent. If the end user is limited to laptops, desktops or standalone servers, the model building team must make decisions on agent behavior aggregation.

b) *Number of agents.* At the heart of every simulation is the question of how many agents are going to be used in the simulation. Answering this question requires balancing multiple competing factors.

   i)   What is the optimum number of agents to use to fully cover all aspects of the model behavior? The ideal design would be to have each element of the model represented by an individual agent. The tradeoff is that this is computationally very expensive.

   ii)  What is the minimum threshold of agents that can be aggregated and still accurately represent the modeled behavior? In order to reduce the hardware required (processor, disk space, memory), a design option is to let a single agent represent the behavior of multiple elements of the model. This approach reduces the hardware footprint, but loses some of the granularity of the results.

iii) How complicated is the behavior of each individual agent? In addition to impacting the capacity for emergent behavior in the simulation, having agents with many rules for modeling complicated behavior significantly increases the time required to process each simulation time step (tick). A reduction in simulation execution time directly affects the time to results metric.

c) *Performance tuning.* There are multiple ways to write the software that generates the behavior of an individual agent each tick. It is imperative that the model building team uses best practice software development patterns when designing the agent behavior. Additionally, the performance of the model should be methodically measured and improved to eliminate all unnecessary delays in processing.

3)  How will enhancement and support for the model be provided?

Scope creep is a constant danger faced by any agent based modeling team. Many of the agent system tools that are in use today intentionally make it easy to "squeeze in one more feature" before the final release. The pitfall is that the model becomes unwieldy and the user interface becomes cluttered. And, as with any software product, there are always new features to be added. This leads to the inevitable question of who owns the intellectual heart and soul of the product.

The answer is that both the end users and the development team have a stake in the future of the model. The end user must understand that nothing comes for free. The development team must understand that they cannot operate in a vacuum.

The UNC Charlotte and IntePoint team has been creating modeling and simulation products for the last 9 years. Here are a few suggestions from our experience about creating a sustainable modeling and simulation (M&S) products:

a)  Do not create monolithic models. Divide the functions into discrete modules that can be readily understood by the customer, the current development team, as well as future members of those teams.

b)  Create application programming interfaces (APIs) so that other teams can write new functionality for your product. This allows you to expand the scope easily, allows the customer to have confidence that the model will be supported by others, and allows the model to incorporate knowledge and expertise that does not exist on your team.

c)  Document choices, impacts, and features thoroughly. Everyone associated with the use, extension, and support of the product needs to have a centralized reference that describes what the system does and why.

In summary, adoption and use of the agent-based model depends not only on the design of the agents, but also on a web of non-modeling decisions about how your end user will engage with and leverage the model.   Before coding begins, understand the desired end state that supports your customers' usage habits, time constraints, computational limitations, and long-term expansion and maintenance of the model.

# Chapter 8
# Towards a Characterization and Systematic Evaluation Framework for Theories and Models of Human, Social, Behavioral, and Cultural Processes within Agent-Based Models

Auroop R. Ganguly[1,2,*], Joseph Whitmeyer[3], Olufemi Omitaomu[2], Peter Brecke[5], Mirsad Hadžikadić[3], Paul Gilman[4], Moutaz Khouja[3], Steven Fernandez[2], Christopher Eichelberger[3], Thom McLean[4], Cathy (Yu) Jiao[2], Erin Middleton[2], Ted Carmichael[3], Amar Saric[3], and Min Sun[3]

[1] NortheasternUniversity, Boston, MA, USA
[2] Oak Ridge National Laboratory, Oak Ridge, TN, USA
[3] University of North Carolina, Charlotte, NC, USA
[4] Oak RidgeCenter for Advanced Studies, Oak Ridge, TN, USA
[5] Georgia Institute of Technology, Atlanta, GA, USA

## 8.1 Introduction

An article in Science magazine (Bhattacharjee, 2007) discussed how the U.S. military is interested in enlisting the help of multidisciplinary scientific experts to better understand "how local populations behave in a war zone." The article mentioned the "Human Social Culture Behavior Modeling" (HSBC) program at U.S. Department of Defense and indicated, through a few anecdotal examples, the types of prior research emanating from multidisciplinary fields that may be considered the state of the art. The disparate opinions about the possible value of HSBC models were well captured in the article through the diverging opinions offered by retired military commanders and through the limited success of similar efforts in the past on the one hand versus recent and promising technological developments and data resources (Fig. 8.1) on the other. The critical challenges in systematic evaluation of large-scale social science simulations stem from the inherent multiscale attributes of HSBC processes, models, and theory, as well as from the inadequacy of data and case studies for calibration and validation purposes. The multiscale processes range from psychological profiles of leaders and aggregate crowd behavior to the behavior of institutions or organizations, and of ethnic, geographic, religious, linguistic, and racial groups. The need to adequately handle such processes across scales has spawned a wide range of multiscale social theories, which in turn may be competitive or complementary,

---

* Corresponding author.

M. Hadžikadić (eds.), *Managing Complexity*,
Studies in Computational Intelligence 504,
DOI: 10.1007/978-3-642-39295-5_8, © Springer-Verlag Berlin Heidelberg 2013

and hierarchical or integrated. Also, "surprising" or unusual behavior at one scale may indeed be triggered by minor changes or abnormal behavior at another scale.

Given that validation data are missing, incomplete, or noisy and process understanding is imperfect, the state of the art in model evaluation is expectedly not well developed. Alessa et al. (2006) discussed an "all hands" call for establishing a social science community around the area of "*complexity modeling using agent-based models and cyberinfrastructure*." Quantitative social science theories or theories amenable to quantification in the HSBC domain have seen significant activity in last couple of decades (Carley 1986; Coleman 1990; Opp and Roehl 1990; Opp and Gern 1993; Oliver 1993; Petty et al. 1997; Myers 2000; Ajzen 2001; Ostrom 2007). Many of these theories have been embedded within large-scale simulation systems. Although a variety of methods has been proposed and utilized in the literature (Lewis-Beck et al. 2004), large-scale social simulations have tended to veer toward agent-based modeling (ABM) paradigms in recent years. The ABM approaches themselves have developed along multiple lines. Thus, complex "realistic" agents have been utilized to develop fine-scale models (Silverman et al. 2006) for short-term predictive insights, while models with relatively simple agents (Hudson et al. 2008) have been developed to study emergent behavior. Epstein (2007) describes the generation of artificial societies through ABM, while the Synthetic Environment for Modeling and Simulation (SEAS) described in Chaturvedi et al. (2005) attempts to develop realistic and precise models at country levels. Computational frameworks (Railsback et al., 2006) have been developed for agent-based simulations, from demonstration environments (NetLogo: Tisue and Wilensky 2004) to extensible systems (Collier 2003). However, as discussed by Bonabeau (2002), "*ABM is a mindset rather than a technology*," and as explained therein, thinking of ABM as an alternative to traditional differential equation modeling is wrong, because, "*a set of differential equations, each describing the dynamics of one of the system's constituent units, is an agent-based model*." Bonabeau posits that a synonym of ABM would be microscopic modeling, with macroscopic modeling being the alternative. One example of a macroscopic model is International Futures (IFs), which relies primarily on multivariate statistical analysis and what-if scenario planning with space-time aggregated variables (Hughes 1999). However, even as HSBC models are beginning to explode as a consequence of increased data and computational power, and perhaps enhanced understanding of social processes, commentaries in recent articles like Subrahmanian (2007) make their lack of success in the real world apparent.

Existing methods for the evaluation of theories, models, and systems relevant for HSBC or similar domains rely on the exploration of the hypotheses (or parameter) space and on empirical validation. These methods include active nonlinear tests of complex simulation models (Miller 1998) as well as structural and parametric sensitivity analysis for the evaluation of complex models (Sterman and Rahmandad 2008). These approaches rely on the design of computational experiments (Santner et al. 2003; Husslage et al. 2006) and empirical validation

(Fagiolo et al. 2007; Marks 2008; Windrum et al. 2007). Validation and evaluation in the context of M&S systems for HSBC or similar domains have received some attention from multidisciplinary scientific communities (Vicsek 2002; Tesfatsion and Judd 2006; Bryson et al. 2007).

**Fig. 8.1** Ambient population count at 1 km resolution in Afghanistan obtained from the LandScan program at the Oak Ridge National Laboratory. The availability of precise and accurate geospatial-temporal data in recent years is one of the reasons for optimism in our ability to develop HSBC M&S systems that can produce realistic insights for policy makers. However, prior failures of such endeavors have led to understandable skepticism and imply a need for caution.

Where models are all too imperfect and validation data are inadequate and noisy, traditional calibration and validation approaches are not likely to succeed. Systematic evaluation of models remains useful however, and is perhaps increasing in importance, as decision makers still need to know how to make best use of the available HSBC process understanding, theories and models, as well as how to utilize available data and computational resources as optimally as possible. In these situations, systematic evaluation may have to take the form of characterization of the space of real-world processes and theoretical simulations. The insights gained may have to be qualitative (e.g., tribal loyalties dominate over individual ideologies in a certain region) or quantitative (e.g., based on structural or parametric sensitivity studies). Retrospective and online analysis of observations and simulations may still be useful where such data are available. However, the ability to characterize the real and simulated worlds may ultimately lead to best-fit recommendations on the selection and use of models. Thus, if the ability to produce emergence is desired for a region where populations are known to adopt ideological positions quickly from neighbors within their social network, the type of models selected may be completely different than those selected for a region where short-term prediction is desired for an ideologically driven culture. In this sense, the characteristics of the real-world processes, model simulations, and desired outputs or insights may all drive the evaluation and recommendation strategy. The strategy may rely on automated approaches like Bayesian or response surface methods, analyst-driven approaches based on matching theory with field insights about the population characteristics, or a semi-automated recommendation, where, for example, guided question-answer sessions ultimately result in the model selection process. The selected theories may operate at multiple scales and/or in a hierarchical fashion. The selection of theories may be probabilistic, and the selection process is likely to directly relate to uncertainty formulations, impacts, and risks. Key performance indicators for characterization and systematic evaluation of HSBC theories, models, and systems are described based on a simulation test-bed briefly described in other chapters.

## 8.2   Key Performance Indicators

Key performance indicators (KPIs) and evaluation methodologies have been developed to characterize and evaluate HSBC M&S systems in the following broad areas:

1.  *Causal Dynamics:* Extraction of causal insights and dynamical behavior, obtained by combining structural or parametric sensitivity studies with statistical analysis or data mining of model simulations and observations, where available.
2.  *Emergent Processes:* Characterization and prediction of two types of emergent behavior, the first based on behavioral change and the second on system complexity:

    a.  *Social Behavioral:* Interesting and/or significant macroscale social behavioral patterns explained through microscale social theories, which may also be forewarned using statistical change detection and predicted by social models.
    b.  *System Dynamical:* System characteristics that make conditions favorable to emergence, measured either as the complexity of microscale dynamics or as macroscale signatures of complex nonlinear dynamics.

3.  *Predictability Metrics:* Measures of predictability, obtained as characteristics of a complex nonlinear system like divergence or nonlinear associations among variables.
4.  *Emergence–Predictability Trade-Offs:* Exploration of the possible trade-offs between emergence and predictability (e.g., in nonlinear dynamics, chaos implies a form of emergence and short-term predictability but longer-term loss of predictability; in agent-based models, complex realistic agents are sometimes thought to produce better short-term

predictions even though simpler, interacting agents may model surprises better).

5. ***Dominant Processes:*** The extraction of dominant social processes from potentially massive simulations, as well as possibly limited and noisy observations, through sensitivity analysis and/or data mining approaches like clustering or classification.

6. ***Extreme Behavior:*** A spectrum of statistical distance measures for comparing and contrasting simulations and observations in terms of mean behavior and associations, with a particular focus on the behavior of the extremes and their impacts

7. ***Course-of-Action Analysis:*** The ability of an HSBC theory, model, or system, in addition to the ability of the set of KPIs and methods described earlier, to deliver course-of-action analysis and guidance to end users (e.g., military commanders) with some level of confidence; includes precise predictions, evaluating relatively imperceptible actions in advance that may reduce the likelihood of undesirable emergence later, or the converse, and recommending a set of HSBC theories and models geared to a specific context.

## 8.3   Causal Dynamics and Sensitivity Analysis

The need for sensitivity analysis and for the ability to tease out causal insights from simulations has been discussed in the literature (Sterman and Rahmandad 2008). Sensitivity analysis followed by mining of simulation outputs can uncover the spectrum of causal behavior that individual social theories and their combinations are capable of simulating. Mathematical formulations for sensitivity analysis are relatively well understood (Doubilet et al. 1985; Kleijnen 1998; Saltelli 2004; Hazen and Huang 2006). A promising approach is the generalized likelihood uncertainty estimation (GLUE) concept, originally proposed by Beven and Binley (1992) and described briefly in Saltelli (2004, 173). The GLUE approach can be regarded as a simplified implementation of a Bayesian approach and includes a likelihood-like term that assigns weights to various regions of the parameter space. One advantage is the utilization of the Shannon entropy to keep track of the information content of the weights in the parameter space. The weights assigned to model parameters can be used to develop an ensemble of model outputs at any point in time, thus yielding the output. The weights assigned to multiple competing or complementary models may also be considered as parameters, resembling the multimodel superensemble (Krishnamurti et al. 1999) framework developed for weather forecasting.

## 8.4   Emergent Processes and Emergence Characterizations

Epstein (2007) argues for an operational definition of emergence rather than what he calls "imprecise and possibly self-mystifying terminology of emergence or supervenience." Emergence in complex adaptive systems may appear in unexplained and seemingly inexplicable ways. However, such emergence may have limited value to end users unless it can be interpreted and analyzed for causation. In the context of this chapter, we discuss two forms of emergence that are meaningful and usable but not necessarily comprehensive or mutually exclusive. The first is based on social behavior and the second on system complexity. Walker and Smith (2001) discuss the relative deprivation theory, which forms a cornerstone of social-mobilization-based emergence as discussed later in this paper, while Dessalles et al. (2007) and Boschetti et al. (2005) discuss emergence as a property of system dynamics and complexity.

*Social Behavioral Emergence*

A set of theories for the emergence of social movements is presented (Figure 8.2), which includes a treatment of social mobilization based on relative deprivation.

1. Relative Deprivation Theory: Relative deprivation theory asserts that social actors can feel aggrieved or discontented as a result of feeling deprived compared to some reference point. Egoistic relative deprivation is that felt by individuals; fraternalistic relative deprivation is that felt by members of a group about their group. One secondary step or formulation of relative deprivation theory asserts that social movements can arise when fraternalistic relative deprivation passes some threshold.

2. Forms of Relative Deprivation: The feeling of relative deprivation can occur when (1) the social actor (person, group, perhaps even organization) feels deprived compared to some other peer social actor(s); (2) the social actor feels deprived compared to the actor's circumstances in the past; and (3) the social actor feels deprived compared to its expectation of what its current (or future) circumstances should be.

3. Specific Articulation of Relative Deprivation Theory: An individual experiencing fraternalistic relative deprivation and the accompanying sense of discontent has an increased proclivity to be involved in socially disruptive activity compared to when relative deprivation is absent. Past some threshold of relative deprivation, the individual becomes responsive to and will join socially disruptive behavior (a social movement) if there is one to join. Past yet a higher threshold, some individuals will initiate socially disruptive behavior and thereby increase the likelihood that a social movement will form because of the influence of the initiator on the other members of her or his network (group). A socially disruptive behavior is any behavior that noticeably changes the operation of existing social processes and thus may not be favored by all social groups, especially governments.

4. Network Threshold Theory: This theory is tied to the concept of "tipping point." An individual's decision to participate in a group social behavior depends in part on the activities of those around the individual. (Individuals are embedded in social networks (groups) and respond to the circumstances of those networks.) Beyond some threshold value of the network's (group's) situation, usually in terms of the number of individuals doing some behavior, other individuals will join in doing that behavior. That threshold value

interacts with an individual's proclivity to become involved in a social movement.

5.  Specific Articulation of Network Threshold Theory: An individual will adopt the behavior of others in her or his social network when the number of others in the network doing that behavior passes a threshold. The value of that threshold for individuals is distributed according to some probability distribution.

6.  Instigator Theory: Instigator theory asserts that social mobilization is more likely to occur when an agitator or political leader promotes some behavior by other individuals, and through that agitator's or political leader's influence, the individuals instigate some new behavior. The instigators will advocate a behavior because (1) they are following outside direction, (2) they are socio-political entrepreneurs, and (3) they have a heightened response to relative deprivation.

7.  Specific Articulation of Instigator Theory: The presence of one or more instigators in a group advocating some behavior will, in circumstances that are conducive, stimulate other individuals in the group to do that behavior. In those situations where the instigators respond to a situation of relative deprivation, instigators emerge at different values of relative deprivation. (This value may be distributed according to some probability distribution.)

8.  Articulation of a Simplified Theory of Social Mobilization: Consider a situation with at least two social groups, and (1) the members of at least one of the groups are feeling relative deprivation, (2) there exists within at least one of the aggrieved groups at least one instigator, an individual who responds in a heightened manner to relative deprivation, and (3) the individuals can communicate with each other. A social movement can emerge through the following mechanism: at some point, if the relative deprivation increases because of unfolding events, an instigator crosses her or his threshold for instigation and begins a socially disruptive behavior. The social network may then grow. Thus, if an instigator is in contact with individuals who are past their responsiveness-to-follow threshold, and if the source of their sense of relative deprivation is similar to that of the instigator, then they will adopt the behavior of the instigator.

9.  Stage Set for Social Emergence: If the relative deprivation continues to increase, other individuals pass their thresholds for becoming responsive to the behavior of the instigator(s) in their network and begin the new behavior. This assumes no external actor stops the instigator(s) (repression). If the network or group grows in number, other individuals pass their thresholds for adopting the behavior of their network.

10. A Social Movement Emerges: If those two conditions, especially the second, hold true, the network grows. Eventually the tipping point is reached, and the social movement becomes driven primarily by its network dynamics.

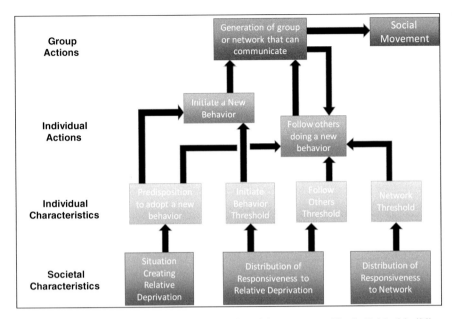

**Fig. 8.2** A diagrammatic view of the theory of social emergence. The individual building blocks and their interactions are presented

*System Dynamical Emergence*

Emergence has been described and categorized qualitatively (de Haan 2006) and somewhat quantitatively (Deguet et al. 2006). Boschetti et al. (2005) mention that although "self-organization may seem to contradict the second law of thermodynamics that captures the tendency of systems to disorder," the "loss of entropy occurs at the macrolevel, while the system dynamics on the microlevel generate increasing disorder." Boschetti et al. (2005) provide an information-theoretic definition of emergence (Figure 8.3) and discuss issues around emergence and predictability from a computational mechanics perspective. Here we investigate the ability to precisely pinpoint microscale dynamics that may be considered sufficiently complex to generate emergence, as well as the ability to extract features from macroscale observables or simulations that are signatures of a system capable of generating emergence. System dynamical emergence includes divergent ensemble runs caused by extreme sensitivity to initial conditions that lead to the presence of strange attractors (see, for example, the Lorenz system: Khan et al. 2005), and also includes cases where relatively simple rules at one scale produce surprisingly ordered behavior at another (e.g., simulating flocks, herds, and schools: Reynolds 1987). Ad hoc definitions of emergence, when a simulation produces seemingly interesting patterns in space and time, may need to be initially developed via spatio-temporal pattern matching and then further investigated to extract dynamical behavior.

**Shannon Entropy for Quantifying Possibility of Emergence at Microscale**
The Shannon entropy of rules' frequency distribution as proposed by Wuensche (1999) has been adapted for agent-based simulations. The input-entropy at time step $t$ becomes:

$$s^t = -\sum_{i=-1,0,1}\left(\frac{Q_i^t}{N}\right) log\left(\frac{Q_i^t}{N}\right)$$

where $N$ is the total number of agents and $Q$ is the number of agents for each $i$ at time $t$. The values of $i$ (-1, 0, +1) are those assigned to the behavior variable, $B$, in the social theories. Wuensche suggests that only complex dynamics exhibit high input-entropy variance.

**Mutual Information for Quantifying Possibility of Emergence at Macroscale**
The mutual information measures the complete dependence, unlike correlations, which are measures of linear associations or rank-based measures that capture only monotonic dependence. The mutual information between agent classes X and Y is the relative entropy between the joint distribution of X and Y [P(X,Y)] and the product distribution P(X) P(Y):

$$I(X;Y) = E_{P(x,y)}\left[log\frac{P(X,Y)}{P(X)P(Y)}\right] = \sum_{k=1}^{m}\sum_{k'=1}^{m}P_{k,k'}\, log\frac{P_{k,k'}}{P_k,P'_k}$$

**Fig. 8.3** Following Boschetti et al. (2005), we use two information theoretic measures for complex dynamics that can lead to emergence: (1) the variance of the input-entropy for the microscale at which the rules operate and (2) the lag-1 mutual information, which provides a measure of macroscale dependence structure. These specific formulations should be viewed as illustrative rather than exclusive or exhaustive.

## 8.5 Predictability Metrics and Measurements

One way to define predictability is to compare retrospectively, or in real-time, with observations and then develop distance measures for the degree of match. However, this approach works only when enough observations are available. If observations are sparse over space and/or time, the HSBC model outputs and states may be validated and updated using approaches like the ensemble Kalman filter (Houtekamer and Mitchell 1998). In situations where observations are inadequate for retrospective analysis or for online updates (e.g., discrete filter formulations) system characterizations may be used to establish bounds on predictability. Thus, linear or nonlinear associations (as determined by, for example, the correlation coefficient or mutual information) between inputs and/or between input-output pairs (where inputs can be a set of input variables and outputs can be a set of simulation variables) would be one type of measure for system predictability. The ability to project the simulations based on simple functional forms, which in turn use current inputs and prior simulation outputs as the functional arguments, may provide an indication of how predictable the system would be with linear or other relatively "simple" nonlinear tools. In cases like the Lorenz series where extreme sensitivity to initial conditions, or "chaos," causes decay in longer-term predictability, the measure of predictability would be based on the sensitivity to initial conditions itself through divergence metrics. Standard methods like Lyapunov exponents and correlation integrals exist to quantify predictability under such conditions e.g., Khan et al. (2005) estimate a few of these metrics from noisy data and small sample size. Nonlinear associations, or information content, among inputs and outputs may provide a way to characterize the predictability of a system. Thus, the mutual information, which extracts the information content concerning one set of variables (e.g., inputs) from another set (outputs), can be a measure of the bounds on predictability.

## 8.6   Emergence versus Predictability Trade-Offs

Within the agent-based model (ABM) community, it has been suggested that ABMs with more realistic and complex agents (as in Silverman et al. 2006 or Chaturvedi et al. 2005), which are inherently more prescriptive, would exhibit more short-term predictability but less ability to model surprising or "emergent" behavior compared to ABMs with simpler individual agents (as in Hudson et al. 2008). These conjectures may be difficult to prove rigorously in the context of large-scale social simulations, ABMs or otherwise, given the differing definitions of emergence, the difficulty of measuring predictability, and the diverse types of models and model complexities. However, we have a clear analogy with complex models in statistics, in which over-parameterization may improve the fit but risks poor generalization. There is also an analogy with nonlinear dynamics, in which additional sensitivity to initial conditions will imply reduced longer-term predictability because the potential space of output grows, even though in the short-term a chaotic system is expected to retain more predictability compared to random noise. Unlike the statistical or nonlinear dynamical analogies, however, balancing the fit and generalization performance, or the chaotic nature of a system with short- and longer-term predictability, may be a significant challenge for HSBC theories, models, and M&S systems. In the latter, the development of equivalents of the information criteria (e.g., the Akaike Information Criteria, or AIC) used to balance model-fit statistics with complexity, or mathematically rigorous formulations to incorporate Occam's razor (MacKay 1995), may not be straightforward. In a similar vein, while the relationship between short- and longer-term predictability, as well as attractor and divergence properties may be well known for chaotic dynamics (Sugihara and May 1990), these insights cannot be immediately generalized to social simulations. Indeed, the trade-offs between emergence and predictability, although seemingly intuitive based on the analogies, cannot be guaranteed as several other factors may overwhelm the process.

Nonetheless, there is significant value to understanding and characterizing the trade-off space, if any, as well as in exploring the predictability and emergence aspects independently.

While several methods may be available for characterizing and measuring emergence (depending on the definition) and predictability, information theoretic approaches provide one common platform to explore the possible trade-offs. In fact, Fig. 8.3 can serve as a surrogate for the emergence versus predictability trade-off, since the microscale complexity measured by the input entropy is a plausible measure of emergence generation, and the mutual information among input-output variable pairs can be a measure of information content and hence predictability. Langton (1991) has explored and interpreted the relationship between input entropy and mutual information in the context of complex dynamical systems.

## 8.7 Extraction of Dominant Processes

The concept of dominant processes as we define it here has, on the one hand, well-established roots within multiple scientific disciplines. On the other hand, the concept presents certain new challenges in the context of the current problem. The system dynamics literature (Legasto et al. 1980; Ogata 1998; Sterman 2000) attempts to model complex, large-scale problems using a combination of relatively simple approaches that comprise feedback loops, causal links, stocks and flows. The simulation system tries to capture the primary underlying drivers, which are often simple at the microlevel, but through nonlinear interactions may be able to model and simulate complex dynamical processes. The selection and combination of the underlying drivers or processes that can best model a complex system must be inferred from observables and matched with simulations during a calibration process. The computational modeling and simulation (M&S) literature (Zeigler et al. 2000; Kuipers 1994) deals with the mathematical and computational issues of efficient M&S implementation. However, the ability to characterize the underlying processes that form the building blocks of the M&S systems must come either from domain theories and insights or from patterns in observed data or from both.

Once again, calibration and validation of the implemented models require a comparison of simulated and observed data patterns, which may in turn be corrupted with noise. In addition, the relevant observed data may have to be gleaned from massive databases or from relatively limited measured samples. The usable portion of the data gathered from massive databases may reveal only incomplete or partial information, as most of the data may not be relevant for the calibration or validation exercise. The simulated data, while potentially unlimited, need to be carefully generated based on statistical experiments that are grounded in domain and data-dictated insights, and with a view to explore the entire range of plausible outcomes. Thus, the ability to extract broader features or patterns from data, both model-simulated and observed, is a key step. The observed data may be noisy or incomplete, and both the observed and simulated data may have to be extracted from massive but possibly irrelevant data. This is where statistical pattern recognition (Fukunaga 1990) and database mining (Han and Kamber 2006) become important. As our discussions may have also illustrated, while extraction of dominant processes for modeling the real world and the inference of predominant features from data are distinct problems, they share a clear and complementary relation. The other angle for dominant process extraction is through data-guided modeling techniques in which the existence of an underlying functional form can be mathematically proven but the actual description is hidden among what are called latent variables or hidden nodes. Just as examples, artificial neural networks (Haykin 1994) and hidden Markov models (Rabiner 1989) fall in this latter category. Although this broad-brush background may be of value as we extend our ideas in this important line of work, the current problem definition and solution strategy are rather narrow and focused, as explained later in this section.

The inference of dominant HSBC processes from the combined use of observations, model simulated outputs, and domain knowledge (e.g., from a human expert) encompasses several aspects of what could be called knowledge discovery, i.e., abduction, analogy, induction, and deduction (Goertzel 1993):

1. Abduction or hypothesis generation would primarily be under the purview of sociologists, political scientists, and economists, but large-scale simulations can support the process by providing an experimental or simulation test-bed (Magnani et al. 1999). The simulation results, and/or the available theories and models, may guide further data collection or discussions with human experts, all of which can help in testing the hypotheses. Thus, let us consider a fabricated but "realistic" situation where the traditional population (e.g., Pashtuns in Afghanistan) acts according to tribal loyalties in a relatively simple-minded fashion, and for them this "process" dominates over other processes like ideology-driven behavior. On the other hand, let us assume that members of an undesirable group (e.g., Arab terrorists) value ideology over all else and have the power to culturally or economically coerce the local Pashtun population to obey their orders. Consider also the possibility that intelligence reports suggest that the Arab terrorist leadership is likely to attempt to infiltrate the Pashtun. Under these circumstances (which may have prevailed just before and after 9/11 in Afghanistan), if new observations begin to suggest to a U.S. military commander or intelligence analyst that the local Pashtuns in a region have started to exhibit more ideology-driven rather than loyalty-driven behavior, a hypothesis may be that Arab terrorist leaders have already infiltrated and have started gaining influence. The commander may then run the simulation with suspected terrorist leadership in place, and if the observations at latter times begin to match the new simulations, then the hypothesis may begin to turn to certitude and suggest the need for action.

2. Analogy or reasoning by similarity (Wojna, 2005) would be extremely useful for generalization of models across geographies and time periods, or conversely, to understand the limits of the ability to generalize. This approach would be important when best-fit model or theory recommendations must be

made on the basis of an understanding of the dominant HSBC processes. Thus, if a model based on tribal loyalty has performed well in Afghanistan, the same model with tuning of parameters and appropriate changes to the input data may work in another region of the globe, say Mongolia, where tribal loyalties also dominate. However, the model may have limited applicability in regions where ideology- and economy-driven behavior may dominate over tribal loyalties (e.g., Indonesia) or where the type of social interactions may be more of the modern urban type rather than tribal (e.g., Russia around Moscow). Thus, if the population in a region is known to have certain characteristics, or if such knowledge can be inferred from a subset of the observations, then this knowledge can be exploited to design best-fit model recommendation strategies. Automated strategies may include analysis of specific traits or outcomes followed by a probabilistic combination (or triggering/suppression) of theories and models. Manual approaches may include creating a catalog of dominant processes across geographies, time, and other relevant dimensions like ethnicity, religious affiliation, or urban sophistry levels. Semi-automated strategies may include question-answer sessions, in which the computer system asks targeted questions to one or more analysts or domain experts and then prescribes or recommends the best-fit model or model combination. Of course, once a "best-fit" model is proposed and selected by analogy, the choice must be continuously validated under new conditions and accordingly modified.

3. Inductive reasoning (e.g., Tenenbaum et al., 2006) may be necessary when dominant processes must be ascribed to populations on the basis of limited observations or even samples that are known to be biased. Thus, it is likely that in a war zone, information about the population within the control of friendly forces may be collected with relative ease, but corresponding information may be difficult or impossible to obtain in enemy territory. In such cases, the simulations for enemy zones may need to run with characteristics induced from a sample of the population that is within friendly control. The other form of induction that may be useful is Bayesian inference, when simulation parameters may need to be initiated based on prior knowledge, and then estimated and continually updated from new data.

4. Deduction: Deductive reasoning (Evans, 2005) forms the cornerstone of modeling and simulation as well as statistics. Hence this form is implicitly utilized in HSBC M&S systems.

While the extraction of dominant HSBC processes is a challenging, for illustrative purposes here we define the process narrowly. We test for dominant processes in two ways. First, we utilize parameter and model sensitivity studies to develop an understanding of causal factors and of how they may influence predominant and minority behavior. This is related to the sensitivity analysis and causal insights section as well. Eventually, we envisage data mining of simulation outputs and approaches using a Markov Chain Monte Carlo (Gamerman and Lopes 2006) for exploring the simulation space. Second, we analyze and utilize observed data—even if limited, incomplete, or noisy—and probabilistically assign the data to the processes that may have generated it. The problem definition is simplified in this case by the fact that we consider only processes that have been encapsulated within the modeling and simulation system. The challenge is to classify the relevant observed data, however sparse or noisy, into one or more dominant process/es, which in turn is/are assumed to be the predominant data-generating mechanism/s. A probabilistic classification (Han and Kamber 2006) would apportion the probability among multiple dominant processes. The classification should pit comparable processes against each other. A close analogue in the machine learning or statistical literature may be the "mixture of experts" approach (Jordan and Jacobs 2001).

## 8.8   Extreme Behavior and Distance Measures

Exploration of the model simulation space requires comparing and contrasting simulation results on the basis of a variety of distance metrics, each of which may measure slightly different characteristics of the difference between the simulations. In situations where observations are available, however noisy or incomplete, a similar set of distance measures would be required to compare observed and simulated data. While detailed observations are expected to limit the calibration and validation of HSBC models (which is what makes the characterization and systematic evaluation described earlier very important), the distance metrics are still necessary for situations where such data exist. The space-time nature of the simulation implies the need for geospatial comparison metrics as developed in, for example, Sabesan et al. (2007). The nonlinear relations among variables and processes may require measures of nonlinear correlations (Khan et al., 2007), while the need to understand extremes and tipping points may require measures based on extremes (Kuhn et al. 2007), anomalies, and threat perceptions that consider threshold exceedances (Sabesan et al. 2007).

The first set of metrics includes aggregate measures like normalized mean squared difference (normalized with the product of the standard deviations of the datasets), fractional area coverage (or the total area of the grids with populations that have certain attributes above predefined threshold values) and grid-based difference measures, with or without transformations like the natural logarithm. These metrics may be displayed using traditional error analysis tools. Spatial plots can be developed to visualize the goodness of fit in space and to detect any obvious, relatively large-scale spatial errors.

The second set of metrics comprises spatial auto- and cross-correlation measures as functions of spatial lags. The spatial cross-correlation metrics can also be interpreted as "spatially aware" measures of difference. The spatial correlations in each direction can be computed by extending the traditional approaches for calculating autocorrelations and correlations used for time-series analysis (Box et al. 1994; Mills 1991), in the context of spatial data. For two spatially distributed variables that are available in spatial grids [e.g., similar to the "lattice" data, as discussed in, for example, Cressie (1993)], the spatial dependence structure between the two variables, as a function of spatial "lags" (distances measured as multiples of grid spacing) in each direction, can be studied by measuring the spatial linear correlation (Pearson product-moment sample correlation) measure.

The third set of metrics is designed to measure the effectiveness of geospatial data in terms of their exceedence above thresholds and hence the corresponding threat or intensity of disruption they may pose in the context of their end use (Murphy 1993). These metrics combine and refine the concepts of "equitable threat scores" and other skill scores used to rank meteorological or climate predictions (Ganguly and Bras 2003). The exceedence-based metrics in Sabesan et al. (2007) are used for evaluating multiple geospatial and geospatial-temporal datasets (e.g., HSBC simulations and relevant observations) rather than prediction skills or signal strengths. The receiver operating characteristic (ROC) curve can be used to visualize the relationship between the false alarm rate and the hit rate. This metric describes the underlying relations in terms of exceeding a threshold and failing to meet the threshold. Measures based on extremes and exceedences, as well as the uncertainties in their estimation, can be related to risk and optimization formulations facilitating end-user decisions.

## 8.9  Course-of-Action Analysis

One ultimate aim of an HSBC M&S system, in the DoD context, may be to consider the diverse political, military, economic, social, infrastructure, and information (PMESII) aspects of the simulation domain and to provide guidance to decision-makers regarding diplomatic, information, military, and economic (DIME) alternatives. Recent advances in social science theory, data and computational resources, and an interdisciplinary toolbox for complex adaptive systems have spawned a new generation of large-scale social simulations. However, our understanding of social processes remains imperfect, and validation data, where available, continue to be incomplete and geographically sparse. On the other hand, there is a need to balance multiple and perhaps disparate objectives. Thus, short-term predictability must be retained along with the ability to anticipate surprises, while simulation systems must be geographically-aware without sacrificing the capability of generalizing across geographies. Satisfying disparate objectives through a spectrum of competing or complementary modeling and simulation approaches requires a detailed understanding of the strengths and weaknesses of each model and simulation system. In addition, quantitative measures are required to characterize relatively abstract yet critical requirements like the ability to produce emergence or model dominant social processes. A quantum leap is needed to move state-of-the-art approaches for evaluation, characterization, and validation to a stage where they can be used to rank models, suggest optimal model combinations, or recommend best-fit simulation strategies. However, developments in computational and data sciences, as well as insights from other complex domains, suggest new possibilities. Best-fit model recommendations leading to precise predictions of predominant social behavior can help devise strategies to win global wars against terror. The ability of these models to anticipate surprises and their triggers may even help win the peace through imperceptible preemptive actions that reduce the possibility of undesirable outcomes in the future.

The predictive insights generated from HSBC M&S systems can be actionable if and only if end users and decision makers can use the insights to help answer questions like the following:

1. Can our understanding of dominant processes, characterization of performance metrics, and evaluation of model performance lead to improved guidance for decision makers for tactical and strategic use? How can M&S systems be positioned to recommend, whether in manual (analyst-driven), semi-automated, or automated (machine-driven) modes, levers to reduce the likelihood of undesirable behavior?
2. Can the possibility of emergence be detected in advance from minor abnormalities in the data or process? Can emergence of undesirable behavior be terminated in the initial stages by relatively small actions taken early, through root-cause analysis of emergence? Can emergence of desirable behavior be similarly encouraged?
3. Can predominant behavior be predicted with sufficient lead time to enable mitigation actions that can thwart unwanted behavior? Can the prediction of expected outcomes, results of root-cause analysis, or predictive insights on emergence be generalized to time periods, societies, and geographical regions other than those for which they were originally generated? Can the points of major upheavals or reversals in public opinions and their root causes be predicted?
4. Can a combination of insights based on emergence and prediction be utilized to design a multipronged decision strategy? Can the modeling system be used by analysts, or can automated methods be designed, to recommend an action strategy? Can the optimal application of such strategies be pinpointed in geography and time with sufficient lead-time to plan and execute necessary actions?

### 8.9.1   A Simulation Test-Bed

A prototype experimental test-bed was designed to evaluate the KPI metrics and methods. A system of simple ordinary differential equations generated complex nonlinear dynamics. Three ABMs were used, and the behavior of each agent was determined by maximizing a utility function. The simplest ABM executed on a small-scale environment considered loyalty and ideology alone. A complex ABM built on a desktop environment considered aggregate data and models. The third ABM utilized high performance computing and fine-resolution data/models.

Three theories of allegiance, specifically legitimacy, representative, and coercion (see Chapter 5), were implemented by assigning appropriate weights to each factor in the utility function. Neighbor interactions were modeled by using two social mobilization theories: (1) social influence and (2) resistance to repression. These are somewhat different treatments of social mobilization compared to the relative deprivation approach described earlier. See Chapter 5 for full presentation of the theoretical justification for these theories and of their implementation in a Cobb-Douglas utility function. Four theories of psychological change, each implemented for change in support for a leader or change in ideology, were developed: socialization, homophily, results-based, and cognitive dissonance. Again, see Chapter 5 for full discussion of the implementation of these theories. Ninety-six combinations of theories resulted from the nine theories (three for allegiance, two for social mobilization, and four for psychological change, where each of the last four can be implemented for support for a leader or change in ideology). The ninety-six combinations were implemented in ABMs along with various heuristics for a case study of Afghanistan. The complex ABM solution was also implemented on a larger-scale computational platform with finer resolution data and models. Although the solutions utilized identical theories, the fine-grained implementations required slightly different heuristics. The rest of this section is cursory; the reader is referred to DARPA Foundation Team (2008) for details.

## 8.10  Computational Implementations

The Lorenz equations are a system of simultaneous ordinary differential equations, which when integrated with a certain set of parameter values demonstrates "chaos," or extreme sensitivity to initial conditions, and several other properties associated with chaos (Weigend and Gershenfeld 1994). The Lorenz system demonstrates that simple systems can produce complex behavior. It has been extensively used by the nonlinear dynamics community to demonstrate basic principles. In our examples, the integrated Lorenz series was often intentionally contaminated with random noise or seasonality (Khan et al. 2005). The purpose was to illustrate certain concepts through a familiar example in nonlinear dynamics, which in turn may be useful for drawing analogies to characterizations of HSBC simulations. The Lorenz equations are given by the following:

$$dx/dt = b\,(y - x);\ dy/dt = -\,xz + rx - y;\ dz/dt = xy - bz$$

Chaotic behavior for certain combinations of parameter values (e.g., $b = 10$; $r = 28$; $b = 8/3$).

A MATLAB (Higham and Higham 2005) system on a desktop personal computer implemented the loyalty and ideology components of the theories of allegiance and the Cobb-Douglas and least squares utility functions. The other components of the utility function were assigned zero weights. This simple implementation allowed a set of followers to be instantiated with different ideological states such that the effect of leader orders on instantaneous behavior could be studied. The purpose of this simple implementation, which can approximate the fully developed M&S system only at a high level, was to provide a mechanism to test new theories prior to implementation as well as to determine the value of the various KPIs through easy-to-understand simulations. In the reported tests, followers were assigned to leaders either randomly or in a very specific way, and the impacts of theoretical settings and leader orders were examined.

A simulation was set up based on NetLogo (Tisue and Wilensky 2004), which is typically used as a demonstration platform for ABMs. We implemented the social science theories and models for contemporary Afghanistan in a personal computer environment. The system considered five types of agents: Afghan government soldiers, coalition forces soldiers, Taliban, leaders, and citizen agents. The citizen agents supportive of the Taliban were called Taliban helpers, whereas citizens who were supportive of soldiers/coalition forces became soldier helpers. The rest were neutral. The country was divided into six regions, each with multiple "patches" in NetLogo. The purpose of the six regions was to allow multiple leaders for the Pashtun tribe, and to allow each Pashtun leader to have a geographically defined area of influence on Pashtun agents. Therefore, the regions apply exclusively to the Pashtun tribe. Other tribes had only one leader each, and those leaders had influence on their agents across the entire country. The data for agents and their attributes were developed in creative ways. For example, opium production was used as a measure of economic prosperity. A variety of heuristics was used for agent behaviors like geographical movements. The total number of agents was limited to a maximum of about 10,000 by the NetLogo environment, which required that the behaviors of citizen agents be modeled at aggregate levels. The data utilized were an aggregate version of the data used for the Oak Ridge Mobile Agent Community (ORMAC) below.

The ORMAC platform (Potok et al. 2003, 2000) was utilized to develop an HSBC M&S system identical to the one in the NetLogo platform but with fine-grained data and with agents at higher resolutions. LandScan population data (Bhaduri et al. 2002) and the relevant geospatial methodologies (Dobson et al. 2000) were used to build a synthetic Afghan population and to geo-locate the 31 million agents. A variety of disparate geospatial sources was utilized to develop and map the agent attributes as well as the theoretical settings. Calibration data were obtained at district levels. The combination of a GIS-based platform with an ABM is by itself a step forward (Brown et al. 2005).

## 8.11   Insights and Limitations

*Insights*

The HSBC M&S system described here demonstrates two major possibilities. First, it shows that multiple social theories at a variety of levels of aggregation can be instantiated in a single model in such a way that they can be compared and combined. Second, it also demonstrates that a variety of data can be incorporated and as needed into the model. The two capabilities are shown by the facts that the demo model runs and that it produces realistic effects. Test cases have verified computational tractability, and a survey of global data sources confirmed the availability of resources to build synthetic populations with the wide range of attributes necessary for the many social theories envisioned for the ultimate system.  By using all-source information and disaggregating primary databases (by means of indicator data sets generated with geospatial science methodologies), synthetic populations may be developed at the individual levels.

The applicability of this methodology was demonstrated in an Afghanistan case study. The ORMAC and the NetLogo simulation systems independently identified coercion theory as most representative of the ground truth observed in Afghanistan, illustrating that HSBC systems may be able to extract the best-fit theory from data. This effort compared and contrasted two different theory-driven, agent-based modeling approaches and explored how each could be implemented and evaluated. A very high-resolution model and a more aggregated model were implemented using the same synthetic population and the same base set of social theories to explore actions that could halt Taliban-inspired terrorist activities.  The aggregated system was able to explore efforts such as force-on-force events and the population's reaction to changes in the security situation. Simulations with agents representing individual social atoms were able to observe the development of small populations of pro-Taliban members. The difference in the results from an aggregate and a finer resolution M&S system was shown. An attempt was also made (DARPA Foundation Team, 2008) to geographically correlate the simulation results with terrorist incidents as recorded in the World-wide Incident Tracking System, which shows that calibration and validation may be a possibility. Utilization of the 31–million-agent Afghan model demonstrated the ability to computationally control (and analytically manipulate) a system with the large number of agents that may be necessary to model populations at the individual level.

## *Limitations*

The insights are confined to relatively simple implementations of theories of allegiance for citizen agents within the utility function framework, with rudimentary adaptive capabilities. Social phenomena at the geographical and temporal scales considered here result from multiple interacting processes, and only a small fraction of these have been addressed in this analysis. Even within the context of the theories we used, the characteristic time scales for the systems are not well studied, and the equilibrium end states or various simulation stabilization criteria are not well understood. Any insights regarding data sufficiency must be treated with caution, especially because the suitability of data for calibration tasks and the relation of data to state variables require detailed examination. The value of fine-grained resolutions and computing power should be objectively tested. Experience in other domains indicates there is often an optimal level of aggregation at which best results are obtained (Chapman and Winker 2005; Fadlalla 2005; Carpenter and Georgakakos 2006; Reed et al. 2004; Done et al. 2004). The optimal aggregation level may depend on the inherent nature of the processes, the validity of the theories and models, and the quality and quantity of input and calibration data. Finally, any insights regarding emergent or surprising behavior must be interpreted with caution (Bonabeau 2007; Epstein 2007) given the subjectivity of the insights and the diverse definitions of emergence.

### 8.11.1 Systematic Evaluation Results

The ordinary differential equation system generated surrogate data, which were used to show how information theoretic measures explore the emergence versus predictability trade-offs. Mutual information among variables related directly to predictability from nonlinear tools, while emergence was obtained by studying the attractor dynamics. The MATLAB-based ABM system had the advantages of little overhead and fast execution times. Thus, sensitivity studies had short execution times leading to the design of a simple test scenario that showcased emergent effects through (1) changes in predominant behavior, (2) precursors of emergence based on analysis and mining of simulated data, and (3) detection of emergent behavior via data abnormalities. The NetLogo-based ABM was used to

- Study the influence of social theory on agent behavior using correlation studies,
- Evaluate microscale dynamics based on input entropy to determine ability to produce emergence,
- Extract macroscale signatures based on the mutual dependence structure to characterize possible emergent behavior,
- Correlate the former dynamics and the signatures to processes and theories,
- Develop dominant attributes by relating pseudo-observations to a contained set of theories that could have generated those outputs, and
- Produce threat scores that measure the closeness of simulation in terms of the ability to balance hits and false alarms in the context of exceedence thresholds.

The results of the ORMAC-based ABM were compared with the results of the NetLogo-based ABM to obtain insights on how the fine-grained data and more resolved process models impacted the final results. The experimental test-bed and the evaluation metrics or methods were developed to show the feasibility of using these approaches in the HSBC domain. In particular, the definition of emergence requires a degree of subjectivity and contextual information. Thus, the definitions or KPI formulations require additional study before they can be generalized to real-world processes or other HSBC models, theories, and domains.

## 8.12 Emergence versus Predictability in Nonlinear Dynamics

The Lorenz X (time series of X obtained upon integration) is corrupted with different signal-to-noise ratios (SNR), and an online support vector regression as proposed by Ma et al. (2003) is used for prediction. The skill (1/MSE, where MSE is the mean squared error) and 1/MI (where MI is the mutual information) between the original noisy and the predicted signals are computed for each SNR, and the results are shown below. The plots (Figure 8.4) indicate that both MSE and MI increase as SNR increases. Mutual information (MI) appears to capture the system predictability. Measures of predictability for small samples and noisy data are in Khan et al. (2007).

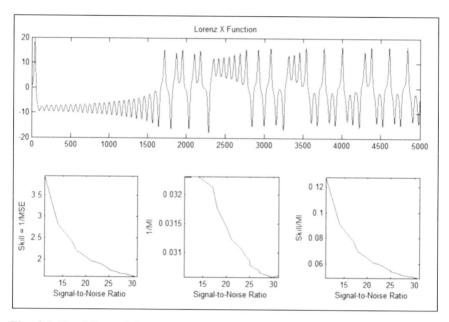

**Fig. 8.4** The MI, an information theoretic measure of nonlinear association among variables, relates to the prediction skill (an inverse measure of predictability) in the Lorenz system, which is contaminated with noise. The bottom plots show, from left to right, the skill, the MI, and both.

The Lorenz X, Y, and Z were compared with random and periodic signals in terms of their dynamics in attractor space. The results (not shown here, but see Weigend and Gershenfeld 1994) illustrate the strange attractor. The ability to distinguish emergent dynamics in the attractor space through information theoretic or fractal measures has been demonstrated (Grebogi et al. 1987; Grassberger and Procaccia 1983). However, whether or to what extent the analogies or the intuitions drawn from nonlinear dynamics remain valid for social science simulations is yet to be demonstrated, specifically in light of model imperfections and data limitations (however, see Judd and Smith 2004). On the other hand, the hypothesis that some of these concepts, which can be clearly demonstrated in the simpler models, would remain valid for more complex simulations cannot be rejected outright and may be a good starting point.

## 8.13   Sensitivity Analysis, Causality and Social Emergence in a Simple HSBC ABM

The MATLAB-based simulation requires selection of four entities. First, either the Cobb-Douglas or Least Squares utility function must be selected. Second, the components of the utility function, or more specifically, the weights attached to the ideology and loyalty components within the utility function (since those were the only two components considered) must be determined. Third and fourth, the impact of ideological positions and the impact of leader orders must be decided. The variable to be studied is the behavior of the citizen agent.

Table 8.1 shows several competing or complementary influences on the final behavioral response.

**Table 8.1** Sensitivity analysis and extraction of causal insights from simulation outputs

Number of citizen agents: 100; Number of leaders: 1
−1 = Pro-Taliban; 0 = Neutral; +1 = Pro-Government
$P, L, I, B$ belong to $\{-1, 0, +1\}$

| Ideological Positioning ($P$) | Selected Utility Function | Leadership Order | Loyalty and Ideology Weights | Behavioral Response |
|---|---|---|---|---|
| {34%, 33%, 33%} | Cobb-Douglas | +1 | 75%; 25% | {0%, 34%, 66%} |
| {34%, 33%, 33%} | Least Squares | +1 | 75%; 25% | {0%, 34%, 66%} |
| {34%, 33%, 33%} | Least Squares | +1 | 25%; 75% | {34%, 33%, 33%} |
| {34%, 33%, 33%} | Cobb-Douglas | +1 | 25%; 75% | {0%, 67%, 33%} |
| {50%, 0%, 50%} | Cobb-Douglas | +1 | 75%; 25% | {0%, 50%, 50%} |
| {50%, 0%, 50%} | Cobb-Douglas | +1 | 25%; 75% | {0%, 67%, 33%} |

A comparison of the first and second rows shows that in certain situations the choice of utility function may not matter. However, comparison of the third and fourth rows shows that the choice of utility function may produce a significant change. Comparison of the first and fourth rows reveals that the choice of weights attached to the different allegiance components (loyalty and ideology in this case) may significantly alter the behavioral response even if the same utility function is used for both simulations. Comparison of the first and fifth rows emphasizes the importance of the initial ideological position. In more complex simulations, the ideological position may itself change over time, but this change is expected to be much slower than the change in behavior or mindshare. A comparison of the last two rows once again illustrates the effect of changed weights in the utility function. Detailed analysis of this nature can help explain observed behavior by suggesting possible root causes. Thus, sensitivity analysis and/or data mining of simulations can quantify the change in behavioral response caused by agent initializations, agent decision attributes, and leadership changes. Once new observations or simulations are acquired or generated, extraction of causal insights and precise predictions can be performed. Observed changes in agent decision-making attributes or interactions may lead to predicted changes in mindshare, while observed or simulated mindshare may be explained by changes in the attributes.

As discussed, detailed sensitivity studies based on a thorough statistical exploration of the simulation output space, and perhaps data mining of the simulation outputs, can yield interesting and useful insights that may be hidden in the simulated data. The hidden knowledge, once discovered, characterizes the simulation system and provides guidelines to the end user. Table 8.2 provides an interesting example.

**Table 8.2** Sensitivity analysis to guide early detection of undesirable behavior

Number of citizen agents: 100; Number of leaders: 1
−1 = Pro-Taliban; 0 = Neutral; +1 = Pro-Government
$P, L, I, B$ belong to $\{-1, 0, +1\}$

| Ideological Positioning (P) | Selected Utility Function | Leadership Order | Loyalty and Ideology Weights | Behavioral Response |
|---|---|---|---|---|
| {50%, 0%, 50%} | Cobb-Douglas | +1 | 75%; 25% | {0%, 50%, 50%} |
| {50%, 0%, 50%} | Cobb-Douglas | 0 | 75%; 25% | {0%, 100%, 0%} |
| {50%, 0%, 50%} | Cobb-Douglas | −1 | 75%; 25% | {50%, 50%, 0%} |

Table 8.2 shows the behavioral states of a population that is ideologically polarized into a pro-government faction and a pro-Taliban faction. We have three cases: (1) a leader is pro-government (top row), (2) the leader is neutral (middle row), and (3) the leader is anti-government or pro-Taliban (bottom row). The utility function is Cobb-Douglas, and the weight applied to loyalty and to ideology is always 0.75 and 0.25, respectively. The corresponding follower behavioral states are 50%-50% neutral and pro-government, 100% neutral, and 50%-50% neutral and pro-Taliban, respectively. The seemingly routine sensitivity analysis hides an interesting characterization of this particular simple HSBC system: even if part of a population has an anti-government ideology, a dominating loyalty component will not allow any anti-government behavior unless the followers obey a rebel leader. This can guide an analyst to look for this type of behavioral response to detect the possibility of rebel leadership, or if this response is found, to suggest the possibility of a rebel leader who may be as yet hidden from public view.

We develop a mock scenario to illustrate how the simple MATLAB-based HSBC system, which has been evaluated based on sensitivity analysis and characterized in terms of causal drivers, can be utilized to detect and forecast social emergence and can lead to course-of-action (COA) analysis. The scenario developed here has a strong resemblance to the theoretical construct for social emergence presented earlier in Section 2.2.1, albeit in a proof-of-concept sense:

1. We have two competing leaders, one pro-government and another pro-Taliban.
2. The followers, or citizen agents, are mostly (90%) pro-government ideologically.
3. A few (10%) followers are ideologically pro-Taliban.
4. A pro-government leader has control over 100% of the tribe initially.
5. The anti-government ("rebel") leader secretly emerges and first controls the 10% of the population who are already ideologically inclined towards Taliban.
6. The influence of the rebel leader extends to the pro-government faction, extremely slowly at first.
7. The influence of the anti-government leader spreads quickly among the social network and rapidly builds momentum.
8. Eventually, the tribe is turned to mostly neutral (predominant behavior), with a significant anti-government faction and only a minority pro-government.

The ideological positions of followers and the orders of leaders are pitted against each other within the context of a utility function, and changes in the resulting behavior of the followers are examined as parameter or model selections vary. A mock test scenario is designed in which a strong leader who backs the government gradually yields influence to a rebel leader. The sensitivity analysis results were able to pinpoint barely perceptible indicators of "emergence," (defined operationally as the growth of influence of the rebel leader) from data abnormalities, and to develop predictions regarding the timing for predominant behavior change as well as possible spread of factors conducive to insurgency.

This identifies windows where small non-kinetic actions at the right time can preclude the need for stronger military action later. The results are depicted in Figure 8.5.

A MATLAB-based example provided earlier in this chapter can also be used to illustrate how dominant processes may become evident from sensitivity analysis. Thus, consider the first and fourth rows of Table 8.1, which show how a switch in the way followers weigh loyalty versus ideology in their individual utility maximization changes predominant behavioral states. In this simple example, assuming all other processes and variables remain constant, the dominant process (the relative influence of loyalty versus ideology) can be easily discerned. In a broader setting, simulated outputs or observations (however incomplete or noisy) may be processed through data mining or information theoretic approaches to characterize dominant social processes that may have generated the observed or simulated data. An example is where simulations generated by a suite of dominant processes are mined and classified into groups, and new observations or simulations are classified (perhaps probabilistically) into one of the groups. This is discussed later.

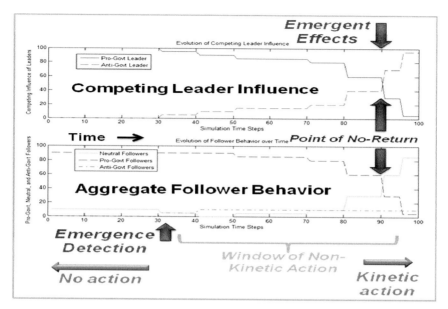

**Fig. 8.5** The presence of pro-Taliban behavior, which can happen in this simulation (Table 8.2) only when a rebel leader is operating, provides advance warning about the possible influence of the leader. In general, this illustrates a case where data abnormalities can provide early warning about emergent behavior in the future. Once the early warning is noted, the simulation can be run to generate the emergent behavior, which in turn can be used to predict the timing and the intensity. Emergence is defined in this case as the point of no return when the predominant behavior changes. This is the point where kinetic action may be the only option. However, prior to this time, a window of non-kinetic action exists when imperceptible action can delay or eliminate the possibility of undesirable emergent behavior.

## 8.14   KPIs from a Complex HSBC ABM on a Demonstration Platform

The NetLogo-based system was used to develop a demonstration scenario, which in turn was adapted from a realistic event in Afghanistan. This realistic scenario was utilized to characterize and evaluate the system and analyze the outputs in a systematic fashion.

*A Storyline and Demo Scenario*
Numerous reports in the press in late 2007 and early 2008 concerned the resurgence of the Taliban in many regions of Afghanistan. One typical article addressed the gradual growth of influence of the Taliban in an area close to Kabul, brought about in part by the intimidation or replacement of the leadership (Parenti 2007). We sought to replicate these events with the simulation model and in so doing, to test the effect of the social influence theory.

A demo scenario was generated on the basis of reported changes in a region near Kabul. The leadership in one region ("Region 2") changes its orders from neutral to pro-Taliban (an effect of increased Taliban activity and control in the area). With social influence turned off (weight 0) the proportion of citizen agents in Region 2 who are pro-Taliban becomes instantly large (in one time step) and completely constant. With social influence turned on (weight 0.4), the proportion of pro-Taliban increases fairly steadily to a rough stability in 15 to 20 time steps. All of this occurs for the legitimacy theory setting of weights. When social influence is turned on, they are scaled back 40%.

With this in mind, we ran a scenario in which the leadership in Region 2 changed its orders from neutral to pro-Taliban in time step 5. We varied the weights of (1) social influence, to see how its setting affected the final equilibrium, the time to change from one equilibrium to another, and the difference between the low and high equilibria; (2) the radius of the effects of influence, either two patches or three; and (3) loyalty, the strongest weight in legitimacy theory. The idea behind these changes was that the two theories combined in the scenario emphasized those two utility function weights (loyalty and influence), so the data would allow a sensitivity analysis of the effects of those parameters separately and together, as well as the effect of one key parameter of social influence, its radius. We ran the simulation twice, once purely with the legitimacy theory of allegiance and once with the legitimacy theory tempered by the social influence theory. Specifically, with only the legitimacy theory, the utility function weights for social influence and repression were set to 0 (utility function weights: loyalty 0.6, coercion 0.1, ideology 0.1, economic welfare 0.2, violence 0, influence 0, repression 0). When social influence was added, the utility function weight for social influence was raised to 0.4, and those for legitimacy theory were reduced by 0.40 (utility function weights: loyalty 0.36, coercion 0.06, ideology 0.06, economic welfare 0.12, violence 0, influence 0.4, repression 0). The psychological change theories were turned off. We replicated the empirical events through one simple change. In the simulation model, the initial order from the leadership in Pashtun Region 2 (the region containing Kabul) is for the population to be neutral. Correspondingly, most of the population in this region is neutral. We mimicked the effects of intimidation or replacement of the leadership by the Taliban by changing the leadership order to pro-Taliban in time step 5. The results were as follows. With the pure legitimacy theory and no social influence, the change in the leadership order had an instantaneous effect: in the next time step, the number of Taliban helpers was several times larger and, in fact, dominant throughout Region 2, and the number of soldier (pro-government) helpers was correspondingly much smaller. After that 1-time-step change, the number of Taliban helpers remained constant. With the social influence theory added, the change in the leadership order had a much more gradual effect. Specifically, the increase in the number of Taliban helpers after one time step was small and concentrated in a small area of Region 2. Over the next approximately 15 time steps, the increase in Taliban helpers spread throughout Region 2 at a fairly constant rate. After 15 to 20 time steps, the number of Taliban helpers remained roughly constant with some fluctuations, at a level somewhat higher than under the pure legitimacy theory. The increase in Taliban helpers did not spread beyond Region 2. Three insights were gained from this exercise: (a) It is possible to replicate at least the broad features of empirical events that fall within the scope of the simulation model; (b) A combination of social theories may yield more realistic results than any single theory. Instantiating the legitimacy theory permitted the change in the orders of the leadership to have effect, but the social influence theory was necessary to add the realism of a gradual spread of that effect; (c) A combination of social theories may produce effects that are not just the sum of the effects from each theory.

*The Key Performance Indicators*

The NetLogo simulation was implemented for 72 time steps using different input weights; each time step was replicated for 41 runs. The following performance indicators were evaluated:

1. **Causal or theoretical insights:** Correlation of outputs with weights assigned to social theories to determine the influence of the individual theories
2. **Emergence based on microscale dynamics:** Inferring system complexity and the likelihood of emergence from microscale dynamics quantified through the input entropy, with an understanding of the causal behavior based on the social theories implemented
3. **Emergence based on macroscale dynamics:** Inferring system complexity and the likelihood of emergence from macroscale signatures quantified through the mutual information dependence structure, with a quantification of the contributing social theories
4. **Emergence versus predictability:** A common platform for exploring the emergence versus predictability trade-off, with normalized measures for both
5. **Dominant process characterization:** Extraction of dominant processes or attributes from data by matching sample data to the simulations generated from underlying theories
6. **Extreme behavior with a threat measure:** A measure that quantifies the contrasts between simulation or observations in terms of extremes through a threat score

The influence of the theoretical settings (in this case, the weights assigned to various components of the utility function) is clearly shown in Figure 8.6 where the difference of the agent classes is seen to be statistically significant as the theoretical parameters change over time.

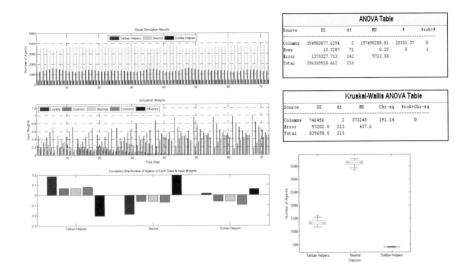

**Fig. 8.6** The average number of agents per class (top left) is compared with the weights assigned to the utility function (middle left) and is correlated with input weights (bottom left). The two-way analysis of variance (ANOVA; top right) and two-way Kruskal-Wallis tests reject the null hypotheses that the mean or medians, respectively, of the classes (columns) are equal across time. The p-values and the box-plots (95% confidence bounds do not overlap) support the rejection.

The typical interpretation of Shannon (or input) entropy is that it specifies the number of bits required to encode (and transmit) the classification of a data item. The entropy is smaller when the data item is more "pure" and is larger when the data item is more "impure." Therefore, entropy has been described as a measure of the rate at which environment appears to produce information. "*The higher the entropy rate, the more information produced, and the more unpredictable the environment appears to be*" (Crutchfield 1994). If entropy is used as a measure of the predictability of classes, then the smaller the class entropy, the more predictable the class would be. For example, if all agents belong to one class, then the entropy is zero, and no bits need to be transmitted because the receiver knows that there is only one outcome (Lee and Xiang 2001); therefore, no uncertainty exists and the class predictability appears much higher. Three classes with the same number of agents will have higher entropy and higher variability because it is more difficult to classify each agent. The input-entropy is a measure of the inherent complexity of the system and hence a necessary condition for complexity-based emergence. Figure 8.7 shows the average input entropy for each class and the sum of the average for all classes at each time step, as well as the variance of the input entropy in each class and for all classes.

As indicated in Figure 8.7, as the input weight for influence increases, the input entropy for each class and all classes decreases. The variance is smallest during the first 20 time steps, is much higher between time steps 20 and 60, and decreases to almost zero at time steps greater than 60.

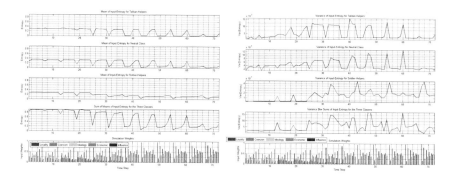

**Fig. 8.7** The input-entropy (left set) and variance of input-entropy (right set) plotted for (top to bottom) anti-, neutral, pro-government, and all agents as a function of the input weights (along the bottom). The low variance at the beginning and end of the simulation period points to relative stability, but the period in between is more volatile.

This distribution may further suggest the presence of phase transition (Langton 1991); however, the direction of the transition is not clear from this experiment. The transitions from an ordered state into a disordered state and vice versa (i.e., phase transitions) are sometimes described as indications of emergence (Langton 1991). Of particular interest is that input-entropy decreases when the input weights for influence and loyalty are 0.0 and at least 0.5, respectively. Minimum input entropy is observed when the weights of influence and loyalty are greater than 0.5 and less than 0.5, respectively. Increased input-entropy gives lower variance. The variance is at a maximum when the input weights for loyalty and influence are approximately 0.4 and 0.6, respectively; and it is at a minimum when the input weights for loyalty and influence are approximately 0.9 and 0.25, respectively. The fact that higher interactions (indicated by relatively larger influence weights) suggest a greater possibility for system dynamical emergence appears intuitive. The mutual information is plotted in Figure 8.8 with respect to two randomly chosen time steps as reference (in this case, the time-steps 1 and 48). The variance of the mutual inputs provides similar insights as in Figure 8.7 earlier, but in terms of macroscale signatures of emergence.

Figure 8.9 shows the lag-1 mutual information, which is a measure of nonlinear dependence among the autoregressive components and hence a measure of predictability in the system. Also shown is the correlation with relative weights assigned to each component in the utility function.

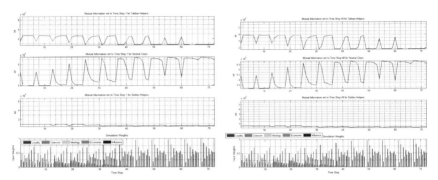

**Fig. 8.8** The mutual information, plotted against two random references, exhibits several features similar to the input entropy in Figure 8.7. The MI starts at some high/low state with much smaller variability, proceeds through a transition with higher variability, then respectively arrives at a new low/high state with much smaller variability. Statistical tests of significance (not shown) confirm that the means of the mutual information are significantly different for each reference, even though the overall trends look similar and are of more interest.

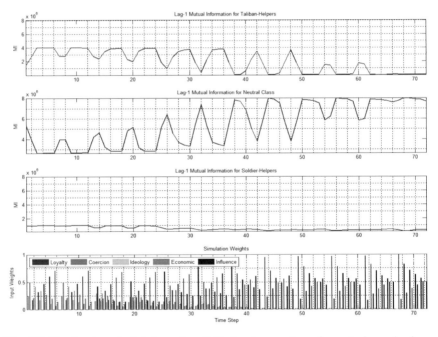

**Fig. 8.9** The lag-1 mutual information (MI) for the number of agents in each class. Evidence of a phase transition is observed for agents of all classes including pro-government (soldier-helpers: SH), but seems more pronounced in anti-government (Taliban-helpers: TH) and neutral (N) agents. A trend with simulation time steps is observed: from high to low MI for TH and SH agents and from low to high MI for N agents. This is similar to Figure 8.8.

Table 8.3 shows the correlation coefficients (CC) between lag-1 MI and the relative weights of each attribute for each agent class. The t-test statistic shows that all the attributes are significant at the 0.05 level except for the CC between loyalty and soldier-helpers.

**Table 8.3** Correlation coefficients between lag-1 MI and relative weights (see Figure 8.9)

| Attributes | Taliban-Helpers (TH) | | Neutral (N) | | Soldier-Helpers (SH) | |
|---|---|---|---|---|---|---|
| | CC | p-value | CC | p-value | CC | p-value |
| Loyalty | 0.3829 | 0.0009 | -0.3498 | 0.0026 | 0.2194 | 0.0641 |
| Coercion | 0.5198 | 0.0000 | -0.5870 | 0.0000 | 0.7790 | 0.0000 |
| Ideology | 0.5198 | 0.0000 | -0.5870 | 0.0000 | 0.7790 | 0.0000 |
| Economic Welfare | 0.5111 | 0.0000 | -0.5806 | 0.0000 | 0.7965 | 0.0000 |
| Inluence | -0.7337 | 0.0000 | 0.7694 | 0.0000 | -0.8527 | 0.0000 |

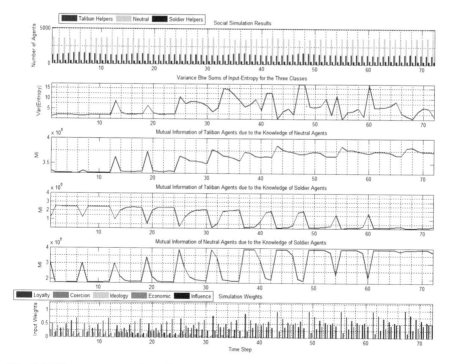

**Fig. 8.10** The average number of agents in each class, the variance between the sum of entropy for the classes, the MI of TH agents due to N agents, the MI of TH agents due to SH agents, the MI of N agents due to SH agents, and input weights. The MI values are plotted on the same scale.

The MI of one class of agents gained by the knowledge of another class of agents is particularly interesting because it measures the mutual dependency between the two classes (i.e., it measures how much the uncertainty about one class is reduced by knowing the number of agents in another class; Cover and Thomas 2006, 19–25). We computed the MI of the number of Taliban-helper (TH) agents due to the knowledge of the number of neutral (N) agents and soldier-helper (SH) agents, respectively, as well as the MI of N agents due to the knowledge of SH agents. It should be noted that MI is symmetric. The plots of these measures are the $3^{rd}$, $4^{th}$, and $5^{th}$ plots in Figure 8.10.

The interesting features in Figure 8.10 are as follows:

1. For these three plots, MI never equals zero. This indicates that the number of agents in each class pair (for example, TH and SH agents) is dependent.
2. At time step 7, after the leadership was switched from pro-Taliban to neutral, we see a sharp increase in MI for N and TH agents ($5^{th}$ and $3^{rd}$ plots in Figure 8.10 respectively) and a sharp decrease in MI for TH agents ($4^{th}$ plots in Figure 8.10); this is not obvious from the variance of input-entropy plot for the three classes ($2^{nd}$ plot in Fig. 8.10).
3. A sharp decrease in MI for TH agents is seen whenever either the input weight for loyalty is 0.0 (regardless of the input weights for other attributes) or the input weights for coercion, ideology, and economic welfare are equal and the input weight for loyalty is 0.5 or less. This is an intriguing observation that may help develop simulation experiments.

The behavior of normalized entropy and MI in a scatter plot has been investigated by Langton (1991). If we interpret entropy as a measure of emergence and MI as a measure of predictability, the scatter plot can be interpreted as a trade-off between predictability and emergence. The scatter plot of normalized entropy versus MI for each class is shown in Figure 8.11. The scatter plots for both TH and SH agents are convex, but the convexity is more pronounced in the plot for SH than in the plot for TH agents. The plot for N agents is concave, however. Consequently, for TH agents, the maximum MI occurs at about the maximum entropy; for N agents, the maximum MI occurs at minimum entropy; and for SH agents, the maximum MI occurs at maximum entropy.

Figure 8.11 again shows evidence of phase transition by way of the gap between the scatter points (see Langton 1991). These gaps appear to classify the points into two or more clusters, which is a trait that has been attributed to the difference in the orders.

- For SH agents, we have three phases: normalized entropy < 0.2 (one state), normalized entropy between 0.3 and 0.8 (transition phase), and normalized entropy > 0.8 (another state).
- For N agents, we have normalized entropy ≤ 0.4, between 0.5 and 0.8, and > 0.8.
- For TH agents, we observe normalized entropy < 0.1, between 0.1 and 0.5, and > 0.5. This class has more gaps than the other two classes. This may be an indication of some useful information that is not clear from this experiment.

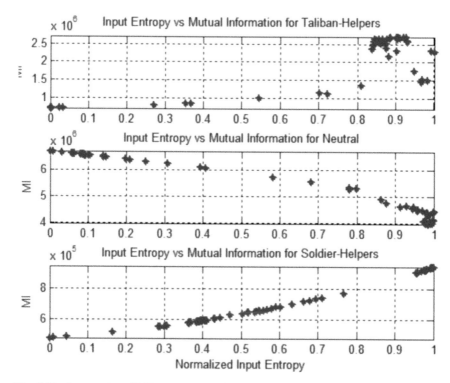

**Fig. 8.11** Scatter plot of MI versus normalized input entropy for each agent class, which may be one way to represent the system-dynamics-driven trade-offs between emergence and predictability.

The extraction of dominant attributes from this particular scenario is difficult as the variation among the number of agents within a class is limited and the effect of any of the five attributes is difficult to distinguish because of the nature of the experiments. Future studies with well-designed experiments can improve upon this aspect. However, Figure 8.12 shows how the input weights and the number of the agents can be visually analyzed.

We note a few interesting points for which we do not yet have full explanations or interpretations. Figure 8.7 indicates that all agent classes have minima for entropy (almost zero) at time steps 43, 55, and 68, all of which had loyalty weights of 0.17 or less and had input weights for coercion, economic welfare, and ideology of 0.02 or less. For all agent classes, the variance approaches a minimum (close to zero in some cases) during the first 20 time steps, then takes a much higher value between time steps 20 and 60, and decreases almost to zero at time steps greater than 60. This distribution may further suggest the presence of phase transition; however, the direction of the transition is not clear from this experiment. Notably, the variance is at a maximum for TH, N, and S agents at time step 40, 60, and 43, respectively.

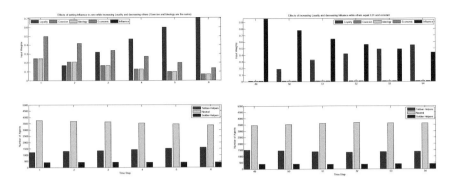

**Fig. 12** The input weights (loyalty, coercion, ideology, economic welfare, influence) at top and the number of agents per class (TH, N, SH) at bottom for times 1–6 (left) and 49–54 (right).

The significance of the input weights tabulated in Table 8.4 may include the fact that for time step 40, the weights for loyalty and influence are the same; for time step 43, the weight for loyalty is 0.00; and for time step 60, the weights for loyalty and influence are almost equal and the weights for other attributes are equal to each other.

**Table 8.4** Input weights at critical time periods

| Time Step | Loyalty | Coercion | Ideology | Economic Welfare | Influence |
|-----------|---------|----------|----------|------------------|-----------|
| **40** | 0.45 | 0.03 | 0.03 | 0.03 | 0.45 |
| **43** | 0.00 | 0.02 | 0.02 | 0.02 | 0.95 |
| **60** | 0.55 | 0.01 | 0.01 | 0.01 | 0.44 |

Distance measures are defined here as metrics that quantify the distance between one simulation output and another or between a simulation output and a set of observations, however sparse or noisy. Thus, correlations, mean squared errors, and information entropy are all useful distance measures. Although simple and commonly used distance measures are useful, here we illustrate a measure that is based on how well the extremes of one align with the extremes of the other. Thus, a mutual exceedence is defined as a "hit" and the reverse is a "miss" and so on. A threat score can then be defined as [hits/(hits + misses + false alarms)]. A class of metrics of this nature has been defined by Sabesan et al. (2007). Usually the measures are employed to compare simulations and observations, but as an illustration, Figure 8.13 shows the threat score when one agent class (SH) is used as a base class.

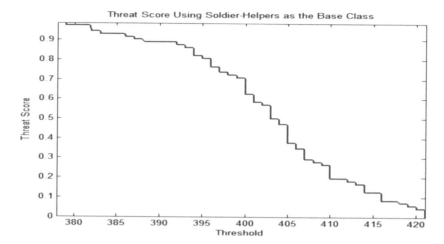

**Fig. 8.13** The threat score computed as a function of exceedence of various thresholds, and plotted as a function of the threshold. Measures like these can be useful to understand how datasets, in this case multiple simulations or observations, differ from each other in the context of extremes. This illustration uses an agent class as a base class and compares it with another agent class.

## 8.15  Comparison of Aggregate and Fine-Resolution HSBC Systems

The ORMAC-based system, with 31 million agents and fine-resolution input data, was compared with the NetLogo-based system, which used a maximum of 10,000 agents and aggregate-level inputs. The initializations were slightly different, with ORMAC using polling data from Gallop for a deterministic initialization and NetLogo using samples from a probability distribution. The geographical and temporal domain was present-day Afghanistan, and the end result was the number of agents with one of three behavioral modes (pro-Taliban, neutral, or pro-government) corresponding to the population mindshare. The social theories embedded in each system were identical, and the test simulations focused on a test of the theories of allegiance described earlier. Each of the two systems was run nine times (Table 8.5), with each run instantiating one of three possible theories of allegiance (legitimacy, L; representative, R; or coercion, C) and one of three time resolutions (3 days, 7 days, or 14 days per "tick," where a tick corresponds to a clock time).

Because the focus was on theories of allegiance, five leader assassinations were introduced to explore the effects of disruptive events (Table 8.6). The events were based on realistic observations in the region. The different simulations provided us with a test set of outputs to demonstrate the value of statistical distance measures. The nine runs are described in Table 8.5.

**Table 8.5** The nine simulation runs

| Run Number | 1 | 2 | 3 | 4 | 5 | 6 | 7 | 8 | 9 |
|---|---|---|---|---|---|---|---|---|---|
| Days / Tick | 3 | 3 | 3 | 7 | 7 | 7 | 14 | 14 | 14 |
| Allegiance Theory | L | R | C | L | R | C | L | R | C |

**Table 8.6** Events in the simulations

|  | Event 1 | Event 2 | Event 3 | Event 4 | Event 5 | End |
|---|---|---|---|---|---|---|
| 3days/tick | 11 | 26 | 27 | 47 | 58 | 68 |
| 7days/tick | 5 | 12 | 12 | 21 | 25 | 35 |
| 14days/tick | 3 | 6 | 6 | 13 | 16 | 26 |

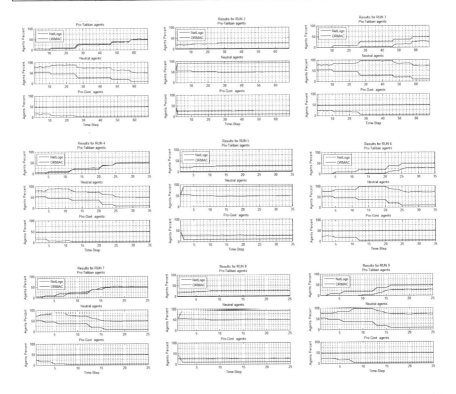

**Fig. 8.14** Nine simulation runs, as shown in Table 8.5, are shown in the nine panels, with the runs arranged sequentially from top to bottom. The outputs generated from ORMAC (blue) and NetLogo (red) correspond to the number of agents exhibiting pro-Taliban (top of each panel), neutral (middle) and pro-government (bottom) behavior.

The aggregate results from the nine simulation runs are shown in Fig. 8.14. The significant bias in the ORMAC versus NetLogo outputs, even at the end points of the simulations where the outputs at successive time steps appear relatively stable, is obvious from the plots. The only exceptions occur during the instantiation of the legitimacy (L) theory and in that case, only for the number of pro-Taliban agents. The fact that simulations differing primarily in their spatial resolutions result in such large relative biases is cause for concern. Comparisons such as these may help improve model outputs, in this case to correct bias errors. The other interesting aspect is that the ORMAC outputs appear more responsive to the disruptive events, while the NetLogo outputs exhibit more random fluctuations but less response to the events.

The difference in the response to events versus random fluctuations is obvious from Fig. 8.15, which shows the first differences, or the differences in the outputs between successive time steps.

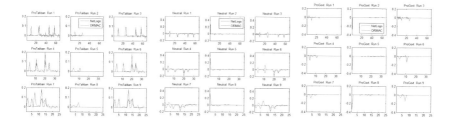

**Fig. 8.15** The first differences for the plots shown in Fig. 8.14, with each of the three panels (from left to right, with nine plots each) showing the nine runs in sequence (left to right then top to bottom). The three panels are, from left to right, for pro-Taliban, neutral, and pro-government classes, respectively, with NetLogo results in red and ORMAC in blue.

The first difference plots in Fig. 8.15 indicate that the response to the events is much clearer in ORMAC, and NetLogo generates more random fluctuations. Although a definitive explanation may not be possible without further investigation, the dominance of random parameters for initialization in NetLogo versus the more deterministic initializations and simulations in ORMAC may be a plausible explanation. In situations where both NetLogo and ORMAC exhibit what appears to be "legitimate" (i.e., occurring at the expected time steps) responses to events, the NetLogo responses seem somewhat damped compared to the ORMAC response. This indeed may be caused by resolution effects, even though there are a few exceptions to this empirical "rule." The lack of any ORMAC response (even the changes in NetLogo appear to be no more than random fluctuations) to the disruptive events when the representative (R) theory is instantiated may be worth noting. The lack of a leader may have less immediate effect on followers when the predominant behavior is representative. However such social explanations must be exercised with care given that the simulation runs appear pretty flat in each case when this theory is implemented (runs 2, 5, and 8). This may suggest an artifact of the specific experimental design. Although the causal explanations offered here are only plausible but not proven unless further simulations are performed, the value of simple metrics (e.g., bias and first differences) together with visual representations may be apparent from the discussions here. Traditional statistical distance measures are shown next. The correlation coefficients (CC) in Table 8.7 further indicate that the increase in the number of days per tick does not significantly improve the dependency between systems with respect to the number of agents in each class.

**Table 8.7** Correlation coefficients between NetLogo and ORMAC outputs

| Run | Days/tick | Theory | Pro-Taliban | Neutral | Pro-Govt |
|-----|-----------|--------|-------------|---------|----------|
| 1 | 3 | L | 0.9750 | 0.8870 | -0.2867 |
| 4 | 7 | L | 0.9776 | 0.9463 | -0.4136 |
| 7 | 14 | L | 0.9610 | 0.9109 | -0.6806 |
| 2 | 3 | R | -0.2505 | -0.2614 | 0.1203 |
| 5 | 7 | R | -0.3966 | -0.4101 | 0.1839 |
| 8 | 14 | R | -0.5601 | -0.6048 | 0.2603 |
| 3 | 3 | C | 0.8479 | 0.1196 | -0.2215 |
| 6 | 7 | C | 0.9361 | 0.4923 | -0.3166 |
| 9 | 14 | C | 0.9479 | 0.5612 | -0.4139 |

**Table 8.8** MSE between NetLogo and ORMAC outputs

| Run | Days/tick | Theory | Pro-Taliban | Neutral | Pro-Govt |
|-----|-----------|--------|-------------|---------|----------|
| 1 | 3 | L | 0.0013 | 0.1804 | 0.1677 |
| 2 | 3 | R | 0.0644 | 0.1648 | 0.0244 |
| 3 | 3 | C | 0.0333 | 0.3132 | 0.1467 |
| 4 | 7 | L | 0.0018 | 0.1567 | 0.1715 |
| 5 | 7 | R | 0.0602 | 0.1697 | 0.0296 |
| 6 | 7 | C | 0.0284 | 0.3089 | 0.1532 |
| 7 | 14 | L | 0.0033 | 0.1521 | 0.1691 |
| 8 | 14 | R | 0.0598 | 0.1596 | 0.0263 |
| 9 | 14 | C | 0.0294 | 0.3159 | 0.1550 |

If the pro-government behavior is ignored (because the ORMAC variations in that category are limited), the correlations are strongest for legitimacy (L) theory, followed by coercion (C). Correlations for representative (R) are negative, confirming the intuition that resolution effects may be very important for that theory. The mean squared errors are rather low (Table 8.8), suggesting that on average for all the simulation time steps, the errors are not too high.

A pair-wise comparison of the correlations between ORMAC and NetLogo runs based on instantiations of different theories of allegiance is shown in Tables 8.9 and 8.10. The results are interesting not just for exploring how the outputs from multiple theories relate to each other, but also for seeing the relations themselves change between the two implementations.

Table 8.9 Correlation coefficients between runs for NetLogo outputs

| Runs | Days/tick | Theories | Pro-Taliban | Neutral | Pro-Govt |
|------|-----------|----------|-------------|---------|----------|
| 1,2 | 3 | L,R | 0.7942 | 0.4615 | 0.6081 |
| 1,3 | 3 | L,C | 0.9166 | 0.4490 | 0.9096 |
| 2,3 | 3 | R,C | 0.5329 | -0.4681 | 0.7442 |
| 4,5 | 7 | L,R | 0.7759 | 0.3778 | 0.5842 |
| 4,6 | 7 | L,C | 0.9149 | 0.5631 | 0.9198 |
| 5,6 | 7 | R,C | 0.5144 | -0.4346 | 0.7077 |
| 7,8 | 14 | L,R | 0.8190 | 0.4502 | 0.6347 |
| 7,9 | 14 | L,C | 0.9378 | 0.6350 | 0.9006 |
| 8,9 | 14 | R,C | 0.6023 | -0.2874 | 0.7761 |

Table 8.10 Correlation coefficients between runs for ORMAC outputs

| Runs | Days/tick | Theories | Pro-Taliban | Neutral | Pro-Govt |
|------|-----------|----------|-------------|---------|----------|
| 1,2 | 3 | L,R | -0.1879 | -0.1879 | -1.0000 |
| 1,3 | 3 | L,C | 0.9996 | 0.9996 | 1.0000 |
| 2,3 | 3 | R,C | -0.1833 | -0.1834 | -1.0000 |
| 4,5 | 7 | L,R | -0.2551 | -0.2551 | -1.0000 |
| 4,6 | 7 | L,C | 0.9996 | 0.9996 | 1.0000 |
| 5,6 | 7 | R,C | -0.2503 | -0.2505 | -1.0000 |
| 7,8 | 14 | L,R | -0.3626 | -0.3627 | -1.0000 |
| 7,9 | 14 | L,C | 0.9998 | 0.9998 | 1.0000 |
| 8,9 | 14 | R,C | -0.3533 | -0.3534 | -1.0000 |

The correlation between the outputs from the legitimacy (L) and coercion (C) theories, as implemented here, is high and appears to be least affected by the simulation granularity (i.e., NetLogo versus ORMAC). However, the relations appear completely altered for any comparison involving the representative (R) theory. Once again, this may be an artifact of the specific simulation, as all the plots (see Fig. 8.14) with this theory appear relatively flat. One other observation is that the change in number of days per tick seems to cause no statistical difference in the dependency between runs with respect to the number of agents in each class.

A detailed comparison of a relatively "lumped" or low-resolution model (e.g., the NetLogo-based implementation) with a relatively more spatially "distributed" or high-resolution model (e.g., the ORMAC-based implementation) typically entails one of two approaches: either aggregate the distributed model outputs to the scales of the lumped model and compare at the aggregate scales, or allocate the lumped model outputs to the scales of the distributed model and compare at the higher resolutions. A comparison based on aggregation may be fairer because the allocation process is not necessarily well defined and may introduce errors. In contrast, aggregation processes are typically well defined; for example when agent counts are considered, a simple sum would almost always be considered appropriate. Indeed, most of the previous comparisons in this section would fall in that category. However, in situations where the end users or stakeholders demand higher resolution outputs, or when the underlying dominant social processes can be best captured at higher resolutions, a comparison at those resolutions may be more appropriate. In such cases, allocations become necessary for comparisons. In the absence of ancillary information, simple methods like equal or area-weighted allocations are probably the only options. In our case, the aggregate "patch" level outputs generated from NetLogo need to be allocated to the finer grids at which data are obtained from the Geographical Information Systems (GIS) and which are ultimately used by the ORMAC-based simulations. The simulation results must be compared at scales that matter to decision-makers (e.g., district levels in Afghanistan). The map for the case study region (Ghazni) with one NetLogo patch, corresponding ORMAC grids, and the Afghan districts (indicated by identification numbers assigned for the purpose of this simulation) is shown in Figure 8.16.

Motivated by the points discussed above, and driven by the need to develop proof-of-concept comparison metrics and methods, we uniformly allocated the aggregate NetLogo outputs to the resolutions of the ORMAC simulations. Specifically, the number of agents for each class and each patch was converted into the corresponding number of agents for each district by multiplying the uniformly distributed number of agents by the number of patches that equal the geographical size of each district. In a sense, this is just an area-weighted allocation strategy. We focused the comparison on "Run 1" (see Table 8.5) and a few districts for illustrative purposes.

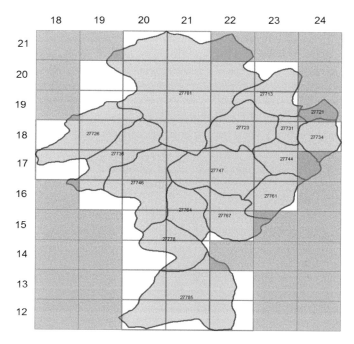

**Fig. 8.16** A map of the Ghazni region in Afghanistan that was used for the case study. The aggregate level NetLogo patch and the finer resolution ORMAC grids are indicated. The district boundaries are marked, and each district is assigned an identification number for the purposes of the simulations.

**Table 8.11** Linear correlation between ORMAC and NetLogo runs at district levels

| RUN | District ID | Pro-Taliban | Neutral | Pro-Govt |
|-----|-------------|-------------|---------|----------|
| 1 | 27713 | 0.9744 | 0.8855 | 0.0000 |
| | 27723 | 0.9759 | 0.8890 | 0.0000 |
| | 27726 | 0.9759 | 0.8897 | 0.0000 |
| | 27731 | 0.9744 | 0.8856 | -0.2867 |
| | 27747 | 0.9740 | 0.8846 | 0.0000 |

Table 8.11 shows high correlations between the ORMAC and NetLogo outputs for pro-Taliban and neutral agents. The zero or negative correlations for pro-government agents may be ignored given that the number of these agents remains relatively constant during the simulation time period (see Fig. 8.14, first panel, bottom plot). The high correlation at a district level indicates that a simple allocation was able to handle the resolution effect in terms of the correlation measure. However, these results should be generalized with caution and only if further experiments are confirmatory. In addition, we note that correlation is one

measure that captures linear associations among data fluctuations, but more nuanced differences may not be obvious by looking at this measure alone. Thus, the effects of resolution are clearly seen in the district-level agent distributions over time, as shown in Figure 8.17. In fact, Figure 8.17 clearly shows that both the time at which predominant behavior changes (e.g., from dominant pro-Taliban to dominant neutral) and the magnitude of the change can be significantly impacted by the simulation and data resolutions (i.e., the ORMAC versus NetLogo simulations). Indeed, the final agent distributions are also significantly affected.

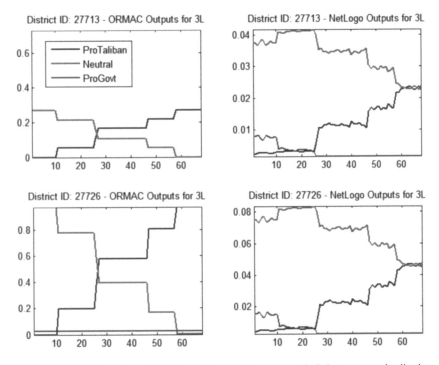

**Fig. 8.17** Distribution of agent fractions (the numbers are scaled) in two sample districts (see Table 8.11) from Run 1 (see Table 8.5). The effects of finer data and model resolutions (i.e., the differences between NetLogo and ORMAC outputs) are clearly seen in the overall magnitudes, the time when predominant behavior changes, and the magnitude of the change. The similarity to Fig. 8.5 is noted.

The similarity of Figures 8.17 & 8.5 is interesting because : (1) the possibility of interpretating Figure 8.17 or similar plots in terms of "social emergence," with links to COA analysis, as in Figure 8.5; (2) the possibility of obtaining equivalent representations of interpretable and useful emergence from disparate HSBC simulation strategies. For the purposes of this chapter, we prefer to leave these as open areas for future research.

### 8.15.1   Distance Measures for Noisy or Limited Observations

Distance measures are routinely used to compare multiple simulations with each other and with observed data. However, the added complexity in HSBC M&S is that the models are more imperfect than usual, dependence across geographies and among entities may be manifest in different ways, and the effects of random noise or randomness in simulations as well as nonlinearities and thresholds may compound the difficulties associated with incomplete observations. In the previous section, we developed or utilized distance measures to compare the NetLogo and ORMAC-based simulations. Distance measures may include bias and mean-squared errors, linear correlation, rank-based correlation measures like the Kendall's Tau, and MI-based nonlinear correlations (Khan et al. 2007). In addition, quantile correlations (e.g., correlations among percentiles) and "threat scores" (Sabesan et al. 2007) have been utilized to explore correlations among extremes (Kuhn et al. 2007), or among low probability but high impact events. Despite considerable progress in the development of data resources to drive HSBC models, observations relevant for calibration and validation remain noisy and incomplete. Thus, the applicability of traditional statistical distance measures remains severely limited. Figure 8.18 illustrates the complexity of the problem.

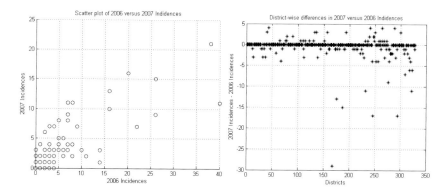

**Fig. 8.18** Scatter plot of 2006 vs. 2007 terrorist incidences, and a district-wise difference plot of the year-to-year differences in the number of incidences, in a region of Afghanistan. The apparent lack of linear associations between the yearly data, as well as the lack of evident patterns in the errors, may be noted. The temporal resolution covering a year, or the geospatial resolution spanning an entire district, may be inadequate to calibrate models that may need to be run on timescales of days to weeks and at the granularity of individual social actors or organizations.

## 8.16   Conclusions

This chapter focuses on systematic evaluation of human, social, cultural, and behavioral (HSBC) modeling and simulation (M&S) systems. The experimental or simulation test-bed developed for the purpose (DARPA Foundation Team 2008), as well as the limitations of any real-world insights drawn from the test-bed at this stage, has been described earlier. The systematic evaluation tasks inherit many of those limitations. On the whole, we urge caution before generalizing the insights or conclusions developed in this article to the real world. However, we believe that an important and promising step, albeit small, has been taken toward achieving the ultimate goal.

The purpose of this chapter was to demonstrate, in a preliminary and proof-of-concept fashion, the feasibility of the following in the context of social science simulations:

1. Perform structural and parametric sensitivity analysis for causal analysis and detection or prediction of socially relevant emergence, with the ultimate goal of developing best-fit model and theory selection and recommendation strategies.

2. Characterize emergence types and develop ways to quantify social emergence, as well as metrics to characterize HSBC M&S systems that are capable of generating emergence in terms of microscale evolution or rules or dynamical processes and in terms of macroscale signatures. (We acknowledge that no one definition of emergence is agreed upon, and perhaps such a universal definition is neither necessary nor desirable. However, operational definitions of emergence from multiple considerations remain useful.)

3. Develop trade-offs between emergence and predictability in HSBC M&S systems with the purpose of characterizing such systems and generating recommendations for use based on end user requirements.

4. Develop methods to extract dominant social processes from observations, simulation outputs, and human insights, with the purpose of enhancing understanding of the underlying social phenomena, suggesting best models or model combinations depending on which processes dominate, and producing insights that can be used by military operators (commanders and strategists).

5. Develop distance measures to compare multiple simulation results, as well as to compare simulations with observations (even when such observations are noisy, sparse, partial, or incomplete), with the goal of evaluating performance of HSBC systems in terms of modeling predominant behavior and processes, extreme values, nonlinear and rare processes, tipping points, and surprising behavior.

The area of HSBC M&S suffers from models that are poorly understood (relative to models for most physical or natural systems) and data that are inherently noisy, sparse, and incomplete. Thus, validation takes on the form of characterization and systematic evaluation, with the ultimate aim of providing value to end users and stakeholders, in this case military commanders. The results presented here take a first step in this challenging direction.

**Credits and Acknowledgments.** The first author and several of the co-authors were employed by the Oak Ridge National Laboratory (ORNL), managed by UT Battelle, LLC, for the U.S. Department of Energy under Contract DE-AC05-00OR22725.

# Chapter 9
# Lessons Learned from the ACSES Model

Joseph Whitmeyer

University of North Carolina, Charlotte, NC, USA

## 9.1   Introduction

A number of potentially useful lessons emerged from our construction of the
ACSES simulation. We list these, with some elaboration, below. Lessons 1-6
concern how to embed social theories. This was the most novel aspect of our
simulation exercise and this is where we have the most to say about the lessons
learned. Lessons 7 and 8 are about implications of the ACSES model for the
technical construction of such models. Lessons 9-11 concern surprise and
emergence, that is, some of the more interesting outcomes that agent-based
models can produce.

## 9.2   Eleven Lessons

*Lesson 1.* **Theories for human actors in different domains may be best modeled
with different kinds of models.**

The ACSES model incorporates theories for its agents in three different
domains. For citizen agents, it has the choice of allegiance, which aggregates to
the outcome of interest, population allegiance. For fighter agents, there are fight-
related choices, such as whether to fight or to flee. Finally, the model has models
for the adaptation of certain citizen agent attributes relevant to the allegiance
choice models.

The ACSES simulation models *choice of allegiance* as *maximization of a utility
function*. The utilities of different choices are calculated deterministically, and the
agent chooses the behavior with the highest utility. We concluded that a
maximization model is best for embedding social theories relevant to choice of
allegiance for several reasons. It allows the incorporation of any number of
situational variables. It can be sensitive to their values over a continuous range if
appropriate. Most advantageously, most social theories can be modeled with
maximization models for agents at the appropriate level (Lesson 2). Incorporating
new social theories in the simulation model is as easy as adding components—
new quantities that feed into the utility to be maximized.

M. Hadžikadić (eds.), *Managing Complexity*,
Studies in Computational Intelligence 504,
DOI: 10.1007/978-3-642-39295-5_9, © Springer-Verlag Berlin Heidelberg 2013

Consider, for example, our first version of the Cobb-Douglas function, which was the function to be maximized. This model implemented the coercion, legitimacy, and representative followership theories, as well as a continuous range of combinations of these theories, and was as follows:

$$U = (1-L)^{w_L} (1-C)^{w_C} (1-I)^{w_I} (1-E)^{w_E} (1-V)^{w_V} . \qquad (1)$$

As equation (1) shows, the function involved interest in being loyal (L), avoiding coercion (C), being faithful to ideology (I), a better economic situation (E) and avoiding violence (V). When we decided to implement additionally two theories of social influence, we simply added two components to the utility function, interest in conforming to neighbors' behaviors (F) and in avoiding repression (R), a quantity that was lessened the more other citizens were defying the local power. We obtained equation (2)

$$U = (1-L)^{w_L} (1-C)^{w_C} (1-I)^{w_I} (1-E)^{w_E} (1-V)^{w_V} (1-F)^{w_F} (1-R)^{w_R} . (2)$$

There is no problem of having too many components, because setting a component's weight, the $w$-exponent, to 0 turns off the component entirely.

Through adjusting the weights, in fact, the user can combine social theories; a component, and thus the theory or theories that involve that component may be emphasized or de-emphasized by raising or lowering $w$. Alternatively, it is straightforward to implement calibration of the model—i.e., determination of the best configuration of weights, the best combination of theories—using real-world data. We did this also in the ACSES model.

*Fighter behavior* is determined by a *Markov model*, meaning that behaviors are chosen probabilistically by comparing random numbers to cutoffs that depend on the current situation. This was appropriate because our interest was not in theories of combat and the behavior of individuals in combat. We wanted a simple model that generated realistic outcomes, and a Markov model generally is that, for situations in which behaviors and outcomes are fairly predictable.

*Adaptation of certain agent attributes* occurs deterministically and gradually as dictated by the current situation. These are processes that do not involve decision-making and are involuntary, thus a deterministic model seemed most suitable. The different models appear to be alternatives that cannot be blended, thus the ability to combine was not necessary.

### Lesson 2. It is possible to embed theories of actor processes at different levels and speeds to work together in a social simulation.

In the ACSES model, the choice of allegiance model and the adaptation of attributes model are for processes at different levels. The adaptation of attributes is more a micro level, occurring within the agent; the allegiance model is more a macro level as it consists of action by the agents as a unit. To fit our assumption concerning the actual process, the adaptation model operates slowly, with small changes at each time step. Choice of allegiance, in contrast, occurs each time step, and when there is a change, is a large change in a single step. The adaptation models were included specifically to feed into and enhance the choice of behavior models; they concern attributes that feed into components of the allegiance model. The effect is a combinatorial expansion of the social modeling capabilities of the simulation, for all social theories possible under four different adaptation models.

*Lesson 3.* **Most social theories are representable by maximization models.**

Most social theories hinge on behavior—perhaps "mental behavior," i.e., thinking—by actors. Any such theory can be represented, at least approximately, with a maximization model for actors, by including an appropriate set of variables with appropriate weights in the utility function. We might note that this approach will not work for functionalist theories (e.g., Parsons 1951) or for theories that involve a natural selection-like process (e.g., population ecology theories of organizational survival such as Hannan and Freeman (1977)); neither of these kinds of theories depends on contingent behavior by actors at any level. We also should note that most social theories, at least as presented in the social science literature, are not precise enough about the relationships between variables to put the theory directly and unambiguously in mathematical form, which is necessary for incorporation in a computer simulation. We take this up in Lesson 4.

*Lesson 4.* **Formal representations of most social theories, in their current state, will require many additional assumptions and simplifications.**

Formal representation of social theories often requires many guesses and uncertain decisions. The general mechanism of the theory is often straightforward to reproduce algorithmically, but when it comes to specific calculation there is usually much left unspecified in the theory. In such a situation, the best a modeler can do is to make credible assumptions, investigate and report the sensitivity of the model to those assumptions—and call upon the social sciences to develop the theory further (see Lesson 5).

One common uncertainty is that frequently the direction of a functional relationship is specified by the theory, but the functional form almost never is. For example, theories of social influence usually say that it should decrease as proximity decreases but do not say whether it should decrease linearly, as distance squared, inversely to distance, or in some other form. The ACSES model had social influence decrease inversely with distance, but that choice was arbitrary.

Another example concerns functional relationship between the amount of the "good" and its value to a person. The public goods literature notes as functional possibilities: linear (second derivative 0), accelerating (positive second derivative), decelerating (negative second derivative), logistic or S-shaped, or inverse logistic or S-shaped (Ostrom 2002, Whitmeyer 2007). There is little doubt about the *direction* of relationships: a better economic situation is more valued, a better security situation is more valued, and so forth. For many of these components, however, there is little knowledge about even the approximate functional relationship. Where there is knowledge it is often easy to incorporate. For economic goods, it seems likely that a decelerating function is best; for example, it is well-known that losses matter more to people than equivalent gains (Kahneman and Tversky 1979). The Cobb-Douglas function in fact is a decelerating function, for the second partial derivative of utility with respect to a weight (the $w$ exponents) is negative. By assuming the Cobb-Douglas function, therefore, the ACSES model assumes this functional relationships for all the components, not just the economic one.

The variables as well usually require decisions concerning measurement and meaning that may have little guidance. Let us consider again social influence, the basic idea of which is straightforward: the more actors around an actor do something, the more likely that actor is to do it too. As a measure of the "actors around an actor," the ACSES model takes the proportion of all actors taking an ideological position (weighted inversely by their distance). Yet, it could be that neighbors taking the "neutral" position should have less weight because they are often less visible. Or it could be that the effect of a proportion depends on an individual's prior tendencies, such that even a small cardinal number has a positive influence if the individual already has a predilection for that position; this would be similar to the Asch effect found in conformity studies.

Other variables have to be simplified although it is known that reality is more complicated. Consider a different component, security against violence ($V$). This is intended to embody a civilian's undoubtedly usual preference for more security and less violence. Yet clearly there are many kinds of security against violence: personal physical security, physical security of the individual and the individual's close family, security of the individual's territorial domain, security of the individual's community, security of the individual's means of livelihood against damages from violence, etc. We must assume all these can be captured reasonably in a single, one-dimensional variable.

The ACSES model also simplifies its outcome variable for individuals, allegiance. Allegiance takes only three values along a single dimension: pro-Taliban, neutral, and pro-government. Yet even in one dimension attitude certainly exists on a continuum. Moreover, allegiance may be multi-dimensional for some individuals, with, for example, these allegiances competing with allegiance to geographical entities, religions or ethnic groups, or even policies. And of course, this nature of "allegiance" may vary across individuals.

*Lesson 5.* **Putting a social science theory in algorithmic form can illuminate areas where additional work is needed on the theory.**

This follows on from Lesson 4. Many social theories, including those the ACSES simulation uses, have been evaluated and supported with empirical data. Where assumptions are necessary to embed the theories in a social simulation, however, are areas where additional work on the theory to remove some assumptions would strengthen the theory. Our earlier examples of the functional form of an individual's subjective evaluation of the amount of some "good," and of how different kinds of security would figure into an actor's global assessment of "security" are such places.

***Lesson 6.*** **Assessing the explanatory importance of social theories or components of them becomes more difficult the more theories are embedded into the simulation.**

In a simulation model, explanation of outcomes consists of investigating the phase space to determine what variable changes produce the outcome in question. As more theories are embedded into the simulation, this task becomes more difficult in the obvious sense that the phase space has more dimensions and so exploring it thoroughly becomes more laborious. Explanation becomes more difficult also because with the increased number of variables, key variables may turn out to be necessary but not sufficient, or sufficient but not necessary; different combinations of different variables may produce similar effects; and variable combinations may produce outcomes a substantial proportion of the time but not all the time. These "problems" are not necessarily bad, for they may correspond to reality. In fact, a signal advantage of agent-based models over, for example, statistical analysis, is that it allows non-linear effects, feedback, reciprocal causation, and so forth, which are precisely the kinds of processes that will make stating causes difficult. It is undeniable, however, that the results may end up messy and difficult if not impossible to distill into a simple conclusion.

***Lesson 7.*** **It is possible and advantageous to integrate a theory-emphasizing simulation like the ACSES model with empirical data.**

Empirical data were integrated with the ACSES model in several ways. Agents were endowed with characteristics (e.g., ethnic identity) that are empirically representative of the *actual* populations under study. The environment in which they operated, for example, the geography of Afghanistan and the distribution of ethnicity and economic well-being over that geography, was informed by data from publicly available sources. Time series data on violent events in one region over several years were used to calibrate (using the data for the first few years) and then to test (using the data for the later years) the model. There were limitations on the data, of course. Data for some variables were not available (e.g., leader orders), and some were available only at aggregate levels (e.g., economic data, available only concerning some economic goods and only at the regional level). Some assumption were necessary, for example, the economic goods for which we had data could be used to approximate the general economic state of the population. Incorporation of empirical data, however, provides a way of testing theories in a simulation environment that somewhat approximates the real situation and in a realistic theoretical space where nonlinear and feedback processes are possible. Incorporation of empirical data also is a step toward practical, perhaps policy-oriented use of a model.

*Lesson 8.* **For simulations of large populations, it may be reasonable to represent multiple real individuals with a single agent.**

In most of our work with the ACSES model, we used a population of about 10,000 agents to represent a real population of about 31 million people. We did compare some results of the 10,000 agent simulation to those of a simulation that represented the entire large population, i.e., with about 31 million agents. The results of the full-scale model were very similar to those with the scaled-down version. This suggests that for aggregated outcomes such as the one we were interested in (population allegiance), using a single agent to stand for many actual individuals may not affect results and conclusions.

*Lesson 9.* **Some social behavior may emerge as a consequence of rules and attributes although it is not coded for explicitly.**

In the ACSES simulation, the combatants grouped together even though this behavior was not coded explicitly. Rather, it emerge from their stay or flee choices when confronted with other combatants. This is a caution against overprogramming agents (a version of Ockham's Razor). Simple empirical observation of a common behavior does not warrant including it in the agent model. Where theory calls for the agent to perform the behavior, or if explicitly coding turns out to be the only way to obtain a clear empirical fact, then doing so is justified. Otherwise, however, emergence from a more basic, theoretically justifiable model should be given a chance first.

*Lesson 10.* **An agent-based model that allows choice among multiple social theories can generate surprising outcomes.**

The effects of different theories may differ in important ways that are difficult to anticipate. This refers not so much to emergence, although emergence may occur, but simply that in a complicated setting with the situation affected by dynamic processes involving agents operating under different models, it may be impossible to predict the differences in results from different social theories. For example, we mimicked real-world events in one district by having, sequentially, a surge in Taliban fighters, a change of orders from the local leader, a surge of government fighters, and a change back in orders from the local leader. Our principal three social theories, legitimacy, coercion, and representative, produced qualitatively different final outcomes—i.e., widespread pro-government allegiance or essentially neutral allegiance—but it is far from apparent why this should be so. This speaks to the usefulness of simulation models in making predictions from theory, and to the usefulness of a simulation model incorporating multiple, alternative theories. It also means, however, that it is important first to test the implementations of the theories on simple situations for which the effects can be anticipated.

*Lesson 11.* **Verisimilitude, emergence, and surprise are facilitated by including a social influence component.**

By social influence we mean that agents affect each other's properties. This effect is typically positive and typically weighted by geographical proximity. In running the ACSES model, including a social influence component with a substantial weight resulted in several notable differences compared to not including it. With social influence, changes in allegiance usually spread gradually rather than being an instantaneous universal change. Such gradual changes sometimes ultimately spread over a wider geographical area than the area that changed instantly with no social influence. At times, including social influence meant that no change occurred when without it a change did occur, say as a result of a change in a leader's orders. Finally, there were situations where including social influence meant that a change did follow on a change in a leader's orders, but the onset of the gradual change occurred only after a delay of variable length. These effects are realistic, not explicitly programmed, for some, impossible to foresee. Given that social theory supports this process, we conclude that social influence should be included for most uses of the ACSES simulation model and similar models of social processes.

# Chapter 10
# Advancing Social Science through Agent-Based Modeling

Joseph Whitmeyer

University of North Carolina, Charlotte, NC, USA

## 10.1   Introduction

We discuss three issues in this chapter. First, we argue that agent-based models are necessary in social science. This also entails the contention that the more common ways of doing social science—the established methods of statistical analysis and the newer technique of data mining—are insufficient. Second, we advocate building into ABMs in social science the capability of choosing and combining theories or models. We argue that this will improve and accelerate theory advancement. Last, we discuss how to model the individual in ABMs in order to facilitate incorporation of theories choice and their combination.

## 10.2   The Necessity of Agent-Based Simulation Models in Social Science

It may be that the person reading this book does not need convincing that ABMs can be useful in advancing social science. Our claim here, however, is stronger than that. We argue that agent-based simulation models are *necessary*. Just as the complex system of the Earth's climate requires an approach that can embody key features of that complexity, so too are many human social processes complex systems that therefore need a similar approach.

One of the guiding precepts of human civilization since the Enlightenment has been that science facilitates progress. That is, the more we know how things work, the better results we will be able to produce. This is an argument for the importance of scientific theory and against pure empiricism, the method of investigating different solutions until finding one that works well enough. While the principle that theory is important has been broadly accepted, there is by no means a uniform agreement on how important it is, and the pendulum has swung toward theory or empiricism at various times in various disciplines. Often these swings are responsive to the external situations, related to advances in theory on the one hand or the availability of data and research techniques on the other. For

M. Hadžikadić (eds.), *Managing Complexity*,
Studies in Computational Intelligence 504,
DOI: 10.1007/978-3-642-39295-5_10, © Springer-Verlag Berlin Heidelberg 2013

example, discovery of the structure of DNA in the mid-twentieth century and the subsequent advances in molecular biology have strengthened the belief that increased theoretical knowledge will facilitate medical progress, for example in developing drugs.

In recent years, data mining has become a common approach to predicting human activity. Technology has enabled the collection of large quantities of behavioral data—on purchasing behavior, self-expression via social media, etc.—which has stimulated the increased use of this empiricist approach, trolling and probing the data to find patterns that will let marketers predict or even shape behaviors, make policy makers aware of nascent political developments, and so forth. Social scientists, perhaps because their focus generally has been causal explanation rather than prediction, have tended not to make use of data mining. Instead, for the past several decades, the scientific method of choice has been to propose a causal model, in which a dependent variable is determined by independent variables, and test this model by estimating its coefficients with some appropriate data. The models are in the form of a mathematical function, such that the dependent variable or a monotonic transformation of it is a function of a power or perhaps logarithm or exponential of the different independent variables.

Data mining and statistical analysis approaches have been and will continue to be useful. They are inadequate, however, because many social phenomena are characterized by, in a word, complexity. Phenomena involving the interaction of many semi-independent entities are likely to manifest nonlinearities, mutual causation, possibly multiple equilibria or even chaos. The underlying assumptions of the data mining and statistical analysis approaches generally rule out such possibilities.

The features of complexity are especially likely if the entities in the phenomenon in question respond to their individual situations in contingent ways, e.g., depending on their environment, on what other entities around them are doing, and so forth. This makes agent-based simulation models frequently an ideal analytic tool. The diversity in entities can be incorporated explicitly; the sophisticated responses and possible diversity in those responses can be built in; and the simulation process allows for mutual causation and nonlinearities. Probabilistic simulation, such as via Monte Carlo techniques, makes it possible to uncover conditions such as multiple equilibria and chaotical behaviors. Many human social phenomena fit the above description precisely: interaction of many semi-independent individuals who respond to their situations contingently. We can expect, therefore, many human social phenomena to exhibit the same features and, therefore, ABMs to prove useful for understanding them.

We give two examples here. Scholars of social movements (e.g., Tarrow 1998) have noted that social movements tend to follow an irregular cyclical pattern, often in delayed tandem with repressive activity by the state. Carmichael (2010) shows a graph with this pattern for political protests and detentions in South Africa from 1970 to 1985 (from Olzak and Olivier 2005) (see Figure 10.1). This is difficult for a statistical model to analyze; for example, depending on the time point in the cycle, police presence may be excessive compared to the amount of protests or the reverse. In contrast, Carmichael offers an explanation by showing that the exhibited pattern is produced by an ABM with two kinds of agents: citizens, who take on a dissent state from 0 to 1, and government agents who repress dissent. Feedback is built into the model, in that the government will insert more agents the more dissent there is, but this is limited by the government's resources, and the government's resources depend on the number of citizens who are not in dissent.

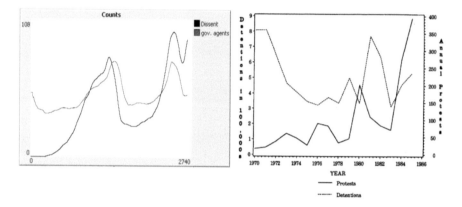

**Fig. 10.1** ABM output of Dissenters and Government Agents (left) and Protest and Police Detentions in South Africa, 1970-1985 (right), from Carmichael (2010)

Another example is a study that hypothesizes that increasing social disorder in a neighborhood increases the level of distrust and feelings of powerlessness among individuals in the neighborhood and supports these hypotheses with statistical analysis (Ross, Mirowsky, and Pribesh 2001). Social psychological theory and studies, however, suggest that individuals' amounts of trust and feelings of efficacy, the opposite of powerlessness, also have positive effects on cooperation. Because more cooperation would mean less disorder, there is reason to expect feedback and reciprocal causation, which, because this occurs in a large population of similar individuals, calls for an ABM. Indeed, an agent-based simulation model of this situation produces hysteresis and two equilibrium values of mistrust and disorder for a range of values of powerlessness (see Figure 10.2) (Whitmeyer 2009). The statistical approach simply would not be able to find this possibility.

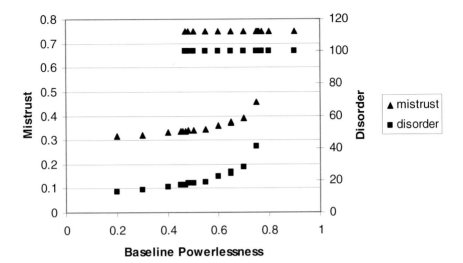

**Fig. 10.2** Equilibrium Mistrust and Disorder for Given Values of Powerlessness, from Agent-Based Simulation Model, from Whitmeyer (2009)

## 10.3    Incorporating Choice and Combination of Theories

Let us describe the "typical" ABM model of a social phenomenon. The modelers have a particular social phenomenon they want to explain. They believe that this phenomenon occurs because of certain key variables that affect individuals' behavioral choices in a certain context of social interaction. The classic example of this is Schelling's model of segregation, which has been expanded and refined many times but the basic idea is the same. Individuals are content only if a certain proportion of their neighbors are like themselves (the social interaction context). They move elsewhere at random if they are discontent (the behavioral choice).

This certainly is a valid and usually useful enterprise. First, it shows that the explanation embodied in the model is plausible, which due to the nonlinearities may be difficult to show without using the model. This may establish the model's explanation as the leading candidate for the correct explanation if no one has been able to explain the outcome before. Or, it may constitute an alternative that challenges other proffered explanations, which are likely to be in the form of some mathematical function tested via statistics, as described above. Second, the user may be able to calibrate the model on empirical data by estimating parameter values, and may be able to generate predictions and test predictions again using empirical data. The user may be able to advance theory by addressing situations for which the model produces clearly wrong predictions. Namely, the user may be able to improve the fit of the outcomes by making reasonable modifications to the theoretical model implemented in the simulation model.

What we believe has been largely missing, however, is appreciation of the fact that an agent-based simulation model does not need to embody only one theory for a given kind of agent. Instead, a single ABM, that is, a single simulation environment populated with a single set of agents, can offer the user a choice or combination of theories. What this means, then, is that we do not have to try to evaluate competing theories through different methods and realizations or wait for a theoretical inadequacy to occur through some empirical problem. Rather we can use repeated runs of our model using the different theories or perhaps combinations of theories to find the best model. This is parallel to a common technique in statistical analysis in which the variables from a number of alternative explanatory hypotheses are incorporated simultaneously into the analysis. The assumption there is that the statistical method will sort out which hypothesis or hypotheses are likely to be supported, through showing "their variables" to have coefficients significantly different from 0. Likewise, an ABM can incorporate multiple theories, and the user can investigate each to see which does better at reproducing empirical outcomes. The user also may be able to combine theories to see if the combination does better than any single theory. We are not criticizing existing uses of simulation models, for supporting the plausibility of a theory is helpful. We believe, however, that being able to go further and support the claim that some theory or combination of theories work better than other theories would be even more helpful.

A second advantage to an ABM that offers a choice among multiple theories or a combination of them is that it is possible that different theories will work better for different populations. In other words, we may want to model some phenomenon as it occurs in different societies or in the same society at different times, for example, citizen allegiance in the presence of an insurgency, the emergence of political or organizational leaders, or the occurrence of demonstrations and protests. The general framework for the phenomenon and the model may be the same in all societies, and even the behavioral model for some kinds of agents may be the same. For example, it may be that in a model of the choice of political or organizational leaders, the same model for potential leaders will work in all societies. For the agents whose aggregate behavior is of primary interest—ordinary people, military fighters, union members, etc.—however, different models may work better in different societies.

Some research in social science suggests that the need for a plurality of models to cover different societies will be true for some phenomena. For example, some scholars who differentiate societies broadly into individualist and collectivist (e.g., Nisbett 2003), meaning that in some societies individuals' orientations are primarily to themselves while in others they are to the group. Other social scientists assert that as societies develop over time, the basic motivations of their members change. This goes back at least to Veblen (1918 [1899]), who claimed that as societies grow more affluent people place increasing value on prestige.

Here, in the interests not only of efficiency but also of facilitating comparison, using the same ABM but allowing the behavioral models for agents to be adjusted for each society would be useful. It would allow, for example, better examination of the claims that people behave fundamentally differently in different societies.

## 10.4    Embodying Theories in ABMs

For embodying theories of individual human behavior in an ABM, we recommend a maximization model. That is, the agent chooses the possible behavior that maximizes some measure. An alternative probabilistic version is that the agent chooses among possible behaviors with probabilities weighted according to the measure. We make a few comments here to clear up possible confusion over such models. First, a maximization model is equivalent to a minimization one. Second, a special kind of maximization model is one where strategy is important. This may call for the inclusion of game theoretic models in the simulation; we discuss this briefly below. Third, a maximization model is not a theory; it is a mathematical tool for representing a theory. In fact, any theory of individual behavior can be represented as a maximization model.

Last, as a special case of the third point, a maximization model is not a theory of "rational" behavior". Discrepancies with "rational"—or perhaps "optimal" would be better—behavior can be captured in a maximization model. To produce bounded rationality or satisficing, for example, it is sufficient to include a cost to acquiring information in the model (Whitmeyer 1998). Human decision-making biases, many of which psychologists have discovered over the years, can be included by designing the maximization model appropriately. For example, people value losses more than gains (Tversky and Kahneman 1981). The Cobb-Douglas and least-squares value functions, two possible forms of a maximization model presented below, possesses that property.

The real content, the theoretic meaning, of a maximization model is in the value function that is maximized. The "maximization" aspect is simply the mechanics of how the model works. In order to implement model choice and combination, we recommend using a function built out of discrete components, each one of which can correspond to a a theory or part of a theory. For example, one form of the Cobb-Douglas function is

$$V = (1 - X_1)^{w_1} (1 - X_2)^{w_2} (1 - X_3)^{w_3} ... (1 - X_n)^{w_n} . \qquad (1)$$

Each $X_i$ ($0 \leq X_i \leq 1$) represents the agent's deficiency, relative to an ideal of 1, in some interest or value to be maximized, corresponding to some theory or theories. The $w_i$s sum to 1 and each $w_i$ indicates the weight of $X_i$ relative to other $X$s. Alternative value functions built out of discrete components are possible, such as the least squares function

$$V = 1 - w_1 X_1^2 - w_2 X_2^2 - w_3 X_3^2 - ... - w_n X_n^2 . \qquad (2)$$

This framework can easily implement theories about how individuals behave. For a theory stressing economic wellbeing, the variable $X_1$, say, could represent the agent's deficiency relative to some ideal economic state—which could change, of course, due to the agent's economic situation changing or the agent's ideal state changing. If prestige were important in some theory, deficiency in prestige could be represented by $X_2$. An insult, then, would increase $X_2$; praise or honor or a promotion would decrease it. Note that the calculation of this $X_2$, and other $X_i$s for other theories, may well be quite involved, depending on the theory and situation being modeled. Other $X_i$s could represent deficiencies in the welfare of those important to an agent, or in some external conditions such as the state of the natural environment, or in the individual's physical or mental wellbeing. In other words, in anything that may orient behavior according to some theory to be implemented. Even a theory that is far from proposing that people find their optimal behaviors and behave rationally can be represented. For example, a theory that postulates that people conform to norms and follow roles can have an $X_i$ represent deviation from the norm.

For the ABM, the components for all desired theories should be included in the value function. A user selects a particular theory, then, by appropriately adjusting the weights $w_i$. In either of the above models, equations (1) or (2), a value $X_i$ is turned off by setting its weight $w_i$ to 0. Thus, if the user wanted a theory that postulated that people are only interested in their economic well-being, measured by $(1 - X_1)$, then the user would set $w_1 = 1$ and all other $w_i$s to 0. Likewise, if the user wanted a theory that postulated that people are only concerned about loss of status, $X_2$, the user would set $w_2 = 1$ and all other $w_i$s to 0. It should be obvious, then, how this allows the user to combine theories; it is done by selecting the weights to be positive for their variables. So, for example, the user could combine the economic and status theories by choosing $w_1$ and $w_2$ with $w_1 + w_2 = 1$, $0 < w_1 < 1$, and $0 < w_2 < 1$. In fact, with good data the user may even want to search for the optimal values of $w_1$ and $w_2$.

Note that in modeling individual agent behavior there are alternatives to the maximization model. The most common one is a Markov model, in which an individual has set of alternative possible behaviors each with a probability assigned, possibly contingent on the situation, and it performs a given behavior with the assigned probability via a random number generator. This method lacks the flexibility of the maximization model and for that reason is not well suited for implementing only simple theories of behavior. It can be useful, however, for obtaining aggregate demographics or for modeling minor agents in a simulation whose behavioral choices are not the focus of investigation. For example, one can set the probability that in a given time step an agent migrates to $p$, or the probability that an agent produces an offspring if it has a certain "age" attribute to $q$. In the ACSES simulation model of citizen allegiance in Afghanistan, for example, we modeled the simple fighting choices of fighter agents—fight, flee, do nothing, move—using Markov models. In contrast, we modeled the allegiance choices of citizens, the behaviors and outcomes in which we were interested theoretically, using a maximization model.

One broad division in the kinds of maximization models to be used is whether we need to consider that individuals behave strategically in the sense of choosing their behavior because it affects others' behavior. Many of the social phenomena modeled by ABMs are mass phenomena; they comprise a fairly homogeneous mass of individuals and are modeled, for the most part, with a homogeneous mass of agents. Any strategic aspect to the behavior of an individual probably is minor and can be ignored. This is due usually to a version of the "tragedy of the commons" problem: an individual's behavior by itself has a negligible affect—but this means that it cannot be strategic. In some scenarios, however, there may be some agents who individually may be able to have an appreciable effect with their actions; for them, therefore, strategic behavior may be possible. In more detail, what this means is that one element in the choice of behavior is likely to be the individual's assessment of the likely responses of others to the behavior in question and how well the individual likes the consequences of those responses. There should be a strategic component, therefore, in the maximization function for such agents in the ABM. Moreover, if there are two or more strategizing agents who strategize with respect to each other, then it is probably appropriate to incorporate a game theory model for the behavior of these agents.

Suppose, for example, we wanted to model the occurrence of demonstrations and protests. We might also be interested in whether the crowd action is violent or nonviolent, and at whom it is directed, such as the government or other groups. The key outcome variables, therefore, will be the proportion of the local population engaging in these behaviors. Members of the population will directly and indirectly influence each other; there may well be threshold effects and bandwagon effects ("momentum"). All of these mean that analysis with an agent-based simulation model is appropriate and promises to be useful. For the key behaviors of likely voters—deciding whether to participate in action directed at some target, and whether to be violent—a maximization model embodying different theories of voter choice will allow the user to choose a preferred theory or to find the best theory or combination of theories.

For the key behaviors of leaders—who can push for such action or against it, for violence or against it, or do nothing--the situation is more complicated. We can assume that one of the high weight values for leaders is obtaining maximal support from the population, but that the leaders have other high weight values such as their personal values, maximizing influence with other leaders, and so forth. The problem with simply applying the maximization framework to the leaders, however, is that, unlike with the behavior of a single member of the population, the behavior of a single leader is likely to affect the behavior of others, that is, other leaders and the population members, which will make the future situation of the leader different from her current one. In other words, leader A's long term optimal strategy may not be what appears maximizing for A in the short term. For example, in order to gain power in the long run, leader A may be better off calling for a popular action A dislikes, perhaps to get an advantage on leader B. Of course a similar logic may hold for leader B as well. In short, intelligent leaders will strategize, which calls for a game theory- application of the maximization models in the ABM.

## 10.5   Final Remarks

We make three final remarks, all of which deserve fuller treatment but not here. First, a question we have not addressed is whether consideration of individual acting entities is always appropriate or the most useful approach in social science. Just as, for many purposes, the properties and behavior of materials and gases may be best treated at an aggregate level rather than the level of individual molecules, it is possible that the same is true for human phenomena. For example, perhaps we can usefully analyze phenomena such as social norms and social institutions without discussing the people who follow or enforce norms or belong to institutions; this notably characterizes the program of Talcott Parsons (e.g., 1951) and his disciples. A different approach that also does without individuals and choosing entities is the population ecology approach to organizations, stemming from the work Hannan and Freeman (1977). Here a Darwinian-type selection process is applied to a diverse population of organizational forms with the result that only certain kinds of organization exist in the end. In contrast, the ABM approach, following many social scientists (e.g., Weber 1978 [1922]; Homans 1964), assumes that it is often advantageous to develop explanations of social phenomena starting from the individual level—but this is an assumption.

Second, we should note one important drawback of ABMs with respect to advancing social science, namely, its lack of transparency. This is true in a minor and potentially correctable way, in the fact that many assumptions frequently are built into the algorithm of the simulation model that are not stated and perhaps not even recognized. In a more significant way, the process by which different parameter settings produce different results may be difficult if not impossible to identify and understand. Thus, while the ABM may, on the one hand, provide an explanation of outcomes, on the other hand it may keep part of that explanation hidden.

Finally, we should mention the possible need to consider important individuals, modeled as important agents, separately. We already noted above that for some phenomena special strategic modeling may be necessary for some agents. Beyond those considerations, however, there may well be important real individuals who have idiosyncratic behavioral tendencies and whose behaviors are too influential for the idiosyncrasies to be disregarded. When, after the fighting was over in the American Revolution, George Washington thwarted a demand by his officers to march on the Congress to demand the arrears in their pay, his choices were due to his personal idiosyncratic preferences and they had tremendous consequences for American and perhaps worldwide democracy (Flexner 1974). The same can be said for his decision, much later, to run for a third presidential term. Models for agents like that must be put in uniquely by the modeler, based on available data.

# Chapter 11
# Additional Problems

Ted Carmichael

University of North Carolina, Charlotte, NC, USA

For the sake of completeness, it may be useful to highlight additional examples, apart from the ACSES simulation, on how agent based modeling can be used to inform policy decisions. In this chapter we will look at two diverse examples of ABM and the types of decision support they can offer. The first is a model of a marine ecosystem; the second is an energy model in the context of an urban environment.

## 11.1 The Marine Ecosystem Model

*Background: Lotka-Volterra*
One of the foundations of ecology dynamics is the set of Lotka-Volterra (LV) equations for predator-prey populations. These equations, while mathematically robust and widely accepted, are general in nature, and limited by the constraints they represent.

First proposed in 1925-1926, the LV equations are a pair of first-order, non-linear differential equations that govern the relationship between two types of interacting species. The equations have periodic solutions, such that an increase in the prey population generates a temporary increase in the predator population, which increases predation levels. Increased predation reverses the growth of the prey population, which in turn reduces the predator population. Once the prey reverses again to a growth phase, the cycle is complete.

Let $x$ equal the prey population and $dx$ equal the rate at which the prey population increases. In the expression of LV, $dx = Axdt$, where $A$ is the growth rate of the prey and $dt$ is the change in time $t$, the prey population x continues to increase exponentially in the absence of some predator.

We define $y$ as the predator population, and $dx = -Bxydt$ as the rate of prey consumption by the predators. Thus the overall growth rate in the prey population is given by combining these two equations:

$$\text{Equation 1:} \quad \frac{dx}{dt} = Ax - Bxy$$

M. Hadžikadić (eds.), *Managing Complexity*,
Studies in Computational Intelligence 504,
DOI: 10.1007/978-3-642-39295-5_11, © Springer-Verlag Berlin Heidelberg 2013

For the predator population, growth is dependent on $C$, the rate of predator removal from the system (either by death or migration), and $D$, the growth rate for predators. Predators therefore also have two equations to express the change in population over time: 1) $dy = -Cydt$, and 2) $dy = Dydt$, which combine to form the following equation:

$$\text{Equation 2: } \frac{dy}{dt} = -Cy + Dxy$$

The solution to these two equations is periodic, with the predator population curve always following the prey curve. Figure 11.1 gives a graphical example of typical LV solution. These dynamics are well understood and have been validated in both computer simulations and real-world studies.

*Generating Lotka Voltera from the Bottom Up*
Following the principles of ABM as outlined elsewhere in this volume, we created a computer simulation model designed to replicate the LV behavior, but by simulating the individual agents, rather than constraining them with the LV equations. (Carmichael, 2010; Carmichael & Hadzikadic, 2013) Further, an ABM instantiation of LV helps to improve our understanding of these equations by allowing for stochastic

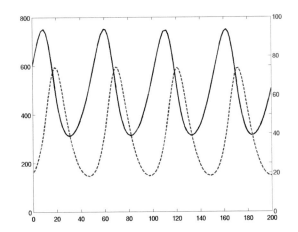

**Fig. 11.1** Graphical illustration of a periodic solution to a typical LV equation, showing the change in population levels for the predator (dashed) and prey (solid) populations.

interactions among the agents. These interactions allow spatial and temporal heterogeneities to arise naturally, as a consequence of these interactions. An ABM simulation also allows us to refine LV by incorporating more flexibility than what is inherent in the equations themselves.

The most obvious refinement on LV in this model is the treatment of the grow rate, for both the prey and the predators: the $A$ coefficient in Equation 1 and the $D$ coefficient in Equation 2. In LV, the prey grow rate $A$ is held constant, and thus assumes an unlimited supply of plankton for the prey. In our model the prey must find food to consume it, and the spatial characteristics of the plankton availability ensure that their consumption rate is variable, not fixed, and that their grow rate is also variable.

The predator growth rate in LV is also constant, explicitly dependent on the availability of prey. Therefore, $Axy$ in Equation 1 and $Dxy$ in Equation 2 are what capture the coupled dynamics between these two equations. (Recall that the $x$ represents the prey population and the $y$ represents

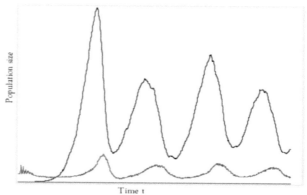

**Fig. 11.2** The prey population (black) and the predator population (red) display the same oscillating behavior as found in Figure 11.3

the predators). As with the prey, the predators must also seek out food, and so their growth rate is also dependent on the *density* of the prey population, not just the absolute number.

Even with these complications, under certain settings the model easily captures the oscillating nature of the predator and prey populations. Figure 11.2 clearly shows the predator population increasing in response to more prey, and decreasing as the prey are consumed. The curves of these population numbers are not as smooth as with the LV equations, indicating the stochastic nature of the ABM.

In the simulation environment, the growth rate of the plankton for the prey can be directly manipulated while the model is running. As the prey's food increases, so does the carrying capacity of the environment. As the model shows, however, this *does not* result in an increase in the number of prey. Rather, the predators also react to a higher food rate, and consume all the extra prey. Thus, all the gains due to an increase in plankton to the low-tropic-level fish accrue to the high-tropic-level predators. Figure 11.3 shows this result as the level of plankton growth is increased from 20% to 30%, and again to 40%. (Specifically, a 20% growth rate of food for the prey indicates a 20 percent chance of food-growth per patch, per simulation time step. The amount of food existent on each patch is recorded as an integer ranging from zero – no food – to ten, the operator-adjusted maximum amount per patch.) In this version the model's parameters are set to minimize the volatility of the two populations so that the changing dynamics between them is more readily visible. This version also incorporates a fourth agent-type, fish eggs, so that the reproduction cycle of the fish is both more realistic and more accessible in the model's settings.

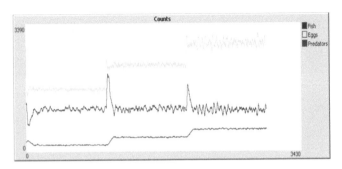

**Fig. 11.3** 3000 simulation time-steps showing population counts at 20%, 30% and 40% food levels (1000 steps per level)

As we can see in Figure 11.3, the prey (red) initially respond to the increase in plankton. (The first third of this graph shows plankton growth at 20%; the middle third has a 30% growth rate; and during the last third growth is set to 40%.) As there is more food available, the prey are more efficient at eating, and thus reproduce at a faster rate. (This is also shown by the increase in the number of fish eggs in yellow.) However, the predators quickly respond in kind, consuming the extra fish and thus increasing their population numbers as well. The predator population remains elevated, while the fish population returns to the original level before the increase in plankton. The fish egg population also remains elevated, indicating that the fish are still eating more and reproducing more, even though that does not translate into a higher population level overall. This is the "stepped pattern of biomass accrual" (Oksanen, et al., 1981; Power, 1992; Brett & Goldman, 1997) and is an important validation of this model.

Our ABM simulation also allows us to closely examine the mean age of each population of agents (Figure 11.4). This information, along with the fish egg population levels,

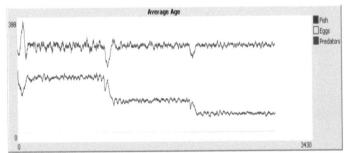

**Fig. 11.4** 3000 simulation time-steps showing mean population age at 20%, 30% and 40% food levels (1000 steps per level); same simulation run as in Figure 11.5

helps to illuminate how the fish population can eat more, and reproduce more, and yet remain at the same level as before we increased their available food. The simulation outputs from Figure 11.4 explain this discrepancy: the fish, though reproducing at a faster rate, do not live as long. This indicates that, collectively, the predators' improved efficiency completely offsets the fishes' improved efficiency.

At first this is due to the higher density of fish, which allows the predators to eat more efficiently and increase their population. When the increasing number of predators overcomes the prey growth, the system reaches a threshold in fish population, and the fish population subsequently declines and returns to a lower density. However, now the predators are more efficient due to *their* higher density, which explains why they stabilize at a higher population, and why the fish stabilize at a lower mean lifetime.

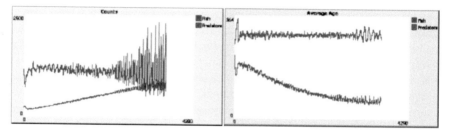

**Fig. 11.5** Increased volatility occurs when the plankton growth rate reaches a certain threshold. Shown is the population levels for predator and prey (left box) and their mean population age (right box). Food supply increased by 1% every 100 ticks, from 20% to 55%. Increased volatility occurs at ~42% food growth.

*Other Validations of the Model*

We have so far shown two validations of this marine ecosystem model: the LV equations, and the stepped pattern of biomass accrual. As stated, the previous experimental design uses parameter settings designed to mute the volatility of the predator and prey populations. But what happens when the food is increased even further, beyond the 40% as found in Figure 11.3 and Figure 11.4? In additional experi-

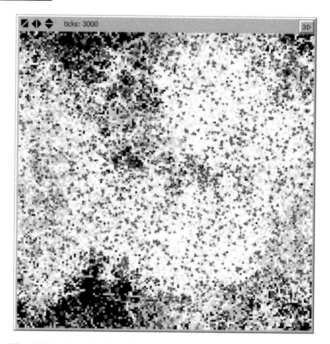

**Fig. 11.6** Spatial clustering of the four populations. Food is allowed to grow in areas (black) where the fish are absent.

ments we increased the plankton growth rate by 1%, from the initial 20% up to 55%. In doing so, we found that there is a phase transition where the prey population suddenly becomes much more volatile. Here this occurs at approximately 42% growth, and is shown in the left plot of Figure 11.5. In the right plot we see that there seems to be some minimum average age, below which the fish cannot reach. This makes sense: if the fish die too young, then they don't have time to reproduce, exasperating the downward pressure on the prey population. Therefore, the fish can only multiply in places where there happens to be an absence of predators. In places where predators are present, the existent fish will quickly be wiped out, allowing the plankton to grow unmolested and forcing the predators to move elsewhere.

In other words, the additional pressure due to a minimum prey age increasingly causes cycles of plenty and famine to emerge across the simulation, spatially. Figure 11.6 illustrates the emergence of this endogenously created heterogeneity.

The emergence of volatility in this model is another important validation that the simulation is able to capture key characteristics of a general food web. The phenomenon represented – The Paradox of Enrichment – is a well-known dynamic across many real-world ecosystems; that is, increasing resources at the lowest tropic level can introduce volatility into the system (Rosenzweig, 1971; Shovonlal & Chattopadhyay, 2007; Scheffer & De Boer, 1995).

Another key phenomenon is the Competitive Exclusion Principle, also known as Gause's Law. It states that two or more species that occupy the same ecological niche cannot co-exist: one will always dominate, and the others will die out. Multiple experiments with this model have shown this principle to hold under a variety of simulation scenarios.

Specifically, multiple species were introduced at the fish tropic level that had no functional differences under the defined simulation interactions. Yet in every case, all but one of these fish species die out. As Gause's Law states, this must be due to some asymmetry between the competing species. Since all of the competing agent-types represented here have the same functional definitions, this outcome must result from a slight advantage due to happenstance during the stochastic agent interactions. Indeed, it is impossible to predict during any single simulation run which species will dominate (Carmichael & Hadzikadic, 2013).

*Informing Public Policy*

These two phenomena – the Paradox of Enrichment and Gause's Law – are important validations of our approach using CAS-based, bottom-up design principles. These, along with the LV equations and the stepped pattern of biomass accrual, show that this model is flexible and realistic enough to generate multiple patterns found in the ecosystem literature.

In terms of policy choices one could look at, for example, fishing practices in various settings. A common method of fishing includes anchoring the nets on the sea floor. As the boats move, these nets are dragged across the seabed, often disrupting years or even decades worth of accumulated systems of growth that are well down the food chain from the target species. In the past, it was assumed that this action would have little effect on the target fish species. However, the clear model results from the stepped pattern of biomass accrual shows that trophic levels that are not directly linked can still be greatly affected. Thus fishing practices such as this may represent two, rather than one, negative pressure events on a particular species: the first due to the direct effects of harvesting, and the second due to destruction of the underlying growth on which such a species may ultimately depend.

Further, worldwide depletion of fishery stocks is of growing concern (Jackson, et al., 2001; Christensen, et al., 2003). This phenomenon – of regional fish population collapse – indicates that there are non-linear dynamics in place that are not yet understood. Yet the ability to experimentally put pieces back into the system to see how it re-forms following collapse offers the potential for better understanding the complex and dynamic relationships for both ecosystem recovery, and the function of ecological systems in general.

The ability to conduct a controlled reintroduction of fish species, in order to address this concern, would require the ability to explore, interactively, the ecology of recovery and the resilience of the larger ecosystem in which such actions take place. Furthermore, the active management of this type of resource depends on a solid understanding of these dynamics to better understand the thresholds at which these systems dramatically change. These factors require simulation tools that take a generative approach, producing emergent patterns of activity through a bottom-up simulation design process. Such a process requires a synergy of field ecology and computing technology to more fully understand emergence in these complex systems.

## 11.2   The Energy Model

*Overview*
We created multiple high-resolution models of energy consumption and generation in the context of an urban environment. The main goal of the overall project was to create a decision-making environment for stakeholders to depict and envision the energy sustainability in the greater Charlotte area. In particular, this environment focused on generating and displaying information about energy use, providing a systematical overview of the energy consumption from multiple simulation systems, and enabling these systems to visually interact with and examine the rich set of simulation outcomes.

Here we focus on one of these models, a simulation of the downtown area, as depicted in Figure 11.7. This model combines commute times with both home and office energy use, by combining approximate guesses as to the intensity of energy use during various home or work activities. Though very simple, and relying on very little specific data, this model already simulates some distinct characteristics of a temporal profile of citywide energy consumption. Specifically, the output plots display "spikes" in energy use over time, as the simulated agents: wake up and prepare for work; arrive at various office buildings downtown; and return home in the evening. Integrated in this model is a simulation of traffic patterns as commuters travel to and from work.

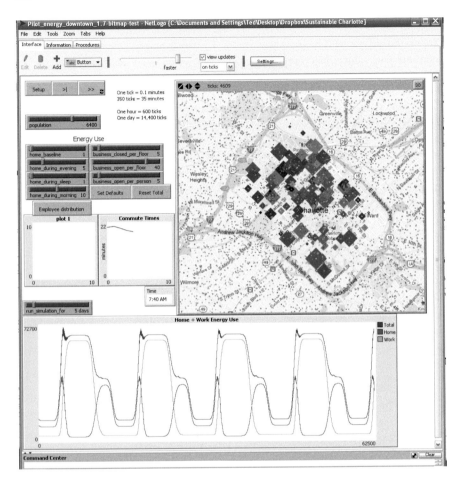

**Fig. 11.7** A partial depiction of the main model's simulation interface. Here one can see the spatial representation of the agents, both buildings and commuters, superimposed on a map of downtown Charlotte. Various controls allow for customized scenario testing by the simulation operator. The main plot depicts home, office, and total energy use over time.

## *Model Specifications*

The simulation model is designed to capture the fundamental dynamic of workers living outside of the center city, and commuting to the center city each day for work. Thus, downtown Charlotte is represented by roughly a rectangle shape approximate to the true center city, and most simulated agents live outside of this area. Buildings are generated here at random in the center city, and houses are also generated at random in the outer areas. (In a later refinement, center city buildings are geo-located based on GIS coordinates, and instantiated with exact size profiles to match the downtown area. While improving verisimilitude of the model, this has very little effect on the overall dynamics described here.) The number of homes is based on the user-selected population size. The number of "floors" per building is also randomly determined, based on the number of workers assigned to that building. The population size is usually sampled as between five and ten thousand; however, there is no limit inherent in the model, other than considerations of run-time efficiency.

Each simulated agent is assigned a home, and a place of work. A random (normal) distribution is used to determine: what time each agent must arrive at work; how long they have to be at work; how long they sleep at night; and how long it takes to get ready for work. Each agent "guesses" initial commute times, and then adjusts those times based on how early or late they arrive, relative to their work start-time.

Each simulation time tick represents 0.1 minutes; thus, 350 ticks is 35 minutes, and one day equals 14,400 ticks. The simulation begins at 12:00 am on the first day, and each agent follows a limited set of activities throughout the day. These are: 1) wake up and prepare for work; 2) travel to its designated work building; 3) travel back home at the end of their work shift; 4) relax at home during the evening; 5) sleep until the next day's schedule begins.

Each of these activities has an assigned energy-use level. These can be adjusted by the user with parameter controls that are built into the simulation. At this stage in the project it was not necessary to use precise energy use; rather, we only required a rough approximation of the *ratios* of energy use during various activities. Therefore the following activities are listed in muW's ("made-up Watts"). A representative example of energy-use activities follows:

- Home baseline, per house: 1 muW.
- At-home, relaxing (before sleep), per agent: 5muW.
- At-home, sleeping, per agent: 1 muW.
- At-home, getting ready for work, per agent: 10 muW.
- Building per floor, when empty: 5 muW.
- Building per floor, with at least one agent: 40 muW.
- Building per person: 5 muW.

With these particular settings, a home that has one person sleeping and one person getting ready for work would use energy at the following rate: 1 muW (baseline) + 1 muW (sleeping agent) + 10 muW (morning agent) = 12 muW per tick. A more detailed example follows:

- A building with five floors.
- Three floors are unoccupied: 3 * 5 muW = 15 muW.
- One floor has two workers: 40 muW (baseline).
- One floor has five workers: 40 muW (baseline).
- Seven workers total: 7 * 5 muW = 35 muW.
- Total: 15 + 40 + 40 + 35 = 130 muW per tick.

It is useful to note that, even though much of the data used for instantiation and rule-based interactions are crude estimates, the resulting energy-use curves are remarkably realistic. For example, our model outputs show a 'spike' in morning energy use as our simulation agents awaken and get ready for work. This is a common feature of real-world urban energy profiles; in fact, it is this spike of maximum demand that is the key concern in energy production management, because it represents the maximum capacity that a produce must be prepared to provide to the system. That is, the energy producer needs to accommodate whatever the maximum level of energy use is, even if only a temporary spike that comes only once per day. Typically, this spike represents the most flexible, and therefore most expensive, mode of energy production.

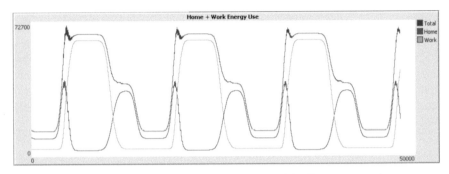

**Fig. 11.8** Detailed image of the main simulation plot. The blue line represents home energy use over time; the green line represents work energy use; and the red line is the sum of home and work.

*Advantages of ABM*

The main advantages of an ABM in this context are two-fold. (1) Complex systems such as this can exhibit multiple levels and forms of non-linear behavior; however, in an ABM, these behaviors can be programmed to arise organically from the system, rather than as top-down inputs or constraints into the model. This allows for a more flexible model. (2) Traditional computer simulations are often referred to as a "black box," due either to the model specifications being hidden in the code, or because the programmed statistical properties of the model are so complicated that they inhibit transparency. A CAS model is inherently transparent. The simulated agents of the model are analogues of real-world agents, whether these agents are people, or ants, or brain cells, etc. Therefore, the assumptions of the model are easily understood as the properties of each agent, and the simple rules of agent-interactions.

An example is useful to illustrate. Suppose we have, rather than an ABM, a mathematical model of cars and drivers, to explore overall system gas use. Statistically, we can program the model to represent the fuel efficiency of various vehicle types with one or more equations. Then we add statistical properties of driving habits, which may also be coupled to vehicle type in a correlated manner. Then we need to add additional equations to represent the density of vehicles on the road at different times of day. Even though such an equation could be as complicated as needed, it would still have difficulty representing both temporal (time of day) and spatial effects, such as congestion spikes that may occur spatially and temporally across areas of the simulated space. These congestion spikes – non-linearities – are often aggregated mathematically over time or space for the sake of simplicity. But there is a difference between, for example, 10 vehicles driving an average of 10 miles each for one hour, vs. 10 vehicles, one of which drives 50 miles in an hour and the remaining nine driving 5 miles in a hour. If these difference matter for representing the real-world analogue of this system, then the equations necessarily become enormously complex, inhibiting transparency.

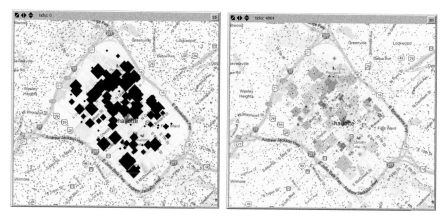

**Fig. 11.9** The main simulation window, showing nighttime energy use (left pane) and early morning commute time with a variable amount of energy use among the downtown buildings (right pane). The downtown buildings are exaggerated in size for easier visualization, and change in brightness based on changes in energy use.

Contrast with an ABM that represents each vehicle as an agent. Each vehicle has a few programmed rules (fuel efficiency, when to drive, where to drive, what speed to travel due to local density) that can easily represent this same system. Each car can have both a fixed (when in use) fuel consumption and a variable (based on speed) fuel consumption. Each simulated vehicle, therefore, can calculate its own variable fuel efficiency during all different times of the day. These calculations can then be aggregated across all the cars – either spatially or temporally (or both) – to produce the known properties of the real-world system.

This inherent flexibility also allows for easy adjustment based on scenarios. Further, the outputs of the ABM do not have to be known *a priori*. Rather, these outputs may be surprising: unanticipated results that emerge with just a few adjustments in agent rules and/or agent interactions. Critically, linear (or non-linear) assumptions are programmed in at the agent level, rather than at the system level. System-level non-linearities, instead, emerge out of the agent interactions. In order to represent such a system with mathematical equations, the scope of all possible agent interactions would have to be known in advance.

*Potential refinements to the model*
Due to the discrete nature of this ABM, additional agent rules and attributes can be easily incorporated, while still preserving transparency. These refinements may

be to simply add verisimilitude to the simulation, or they may be added in order to test various scenarios and answer research questions. Some possible refinements are listed here:

- Vehicle fuel consumption use over time.
- Variable number of agents living in a house.
- Additional housing types, such as apartments and condos.
- A more realistic distribution of workers among the buildings (e.g., based on real-world data if available, rather than estimated as in the current simulation).
- Variability in energy use among buildings based on building age, or based on other measures of building efficiencies.
- Changes in the sub-set of the population living downtown, with the % of those who work downtown commuting without driving.
- Add restaurants, which could introduce economic considerations even in simple ways, such as:
  - o Agents can spend less time getting ready for work, and spend more money to buy breakfast;
  - o "Restaurant" buildings can be added or removed as more or fewer customers frequent them;
  - o These buildings can exhibit spikes during breakfast, lunch, and dinner;
  - o These buildings would have workers that exhibit a different schedule, affecting home energy use as well as commute congestion.
- Add second and third shift workers to some buildings.
- Add weekends (currently, every simulation day is a generic weekday).
- Add light rail.
- "Energy conscious" workers use less energy when they are the only worker on a floor. Thus, individual building profiles may change based on the % of energy conscious workers.
- Add power substations, and represent the energy profiles of each substation individually. This may allow for studies that show, for example, the optimal number and placement of these substations. (Although this could be quite complex, as economic considerations – such as land values and other fixed costs – would have to be known and accounted for.)
- Add entertainment outside of work and home. This would add more categories of energy use, change the dynamics of commutes and traffic, and produce different work schedules for the sub-set of workers employed by this industry.

## *Informing Public Policy*
There are essentially two major types of simulation run experiments that can be conducted in order to inform policy choices: 1) scenarios based on the general characteristics and dynamics of such a model, and 2) scenarios based on specific

characteristics of a well-calibrated model tuned to any target city. Both present certain challenges, but both can also be extremely useful for answering questions about public policy.

An example of the first type of experiment would be to look closely at work times across the simulated agents. Even though the downtown buildings in this model do not represent real and exact office buildings, they do generally represent the size and distribution of such buildings. And so this model can easily explore scenarios of altering work times, either across buildings, across floors in buildings, or – if desired – by building/business type. Because of the non-linearities involved, there is an outsize difference between ten workers arriving on ten different floors at the same time, vs. ten workers arriving simultaneously on the same floor. Further, models of social influence can be tested, in order to give a realistic view of how new policies might spread, fostering multiple types of more sustainable energy use.

Also, a higher degree of downtown living space might well alter the energy dynamics in a positive way. Coupled with a realistic agent-based model of traffic patterns and road congestion, this scenario could also be viewed in terms beyond the power grid, and extend into connected effects on gasoline, traffic, even quality of life. These changes can also be considered in terms of changing the work patterns in the simulation space. For example, more workers living downtown means other services, such as restaurants, emerging in ways that spread out the spikes of energy use within a defined area. Such scenario exploration directly affects the key concern of electricity producers, and can show ways in which density can drastically improve efficiencies across many types of existent infrastructure.

Specific scenarios based on a well-calibrated model, tuned to match a particular city, can also be used as above. However, such a model has additional benefits, in that it can answer questions regarding, for example, where new roads are needed, or better lights for managing traffic patterns. It can explore policy choices that are directly relevant to the types of businesses and industries found in that city, and can help decision makers see the consequences of various choices in a concrete way. The energy producers, for example, could use this type of model to help them decide where to put the next sub-station for optimal results, or to understand how electricity demand curves will change over time. Seasonal scenarios could also be used in a way that matches, for example, the typical weather as summer changes to fall, and then to winter.

This last example is particularly well-suited for an ABM. This is because many effects in the real world are non-linear and path-dependent. Cascading effects can easily be taken into account, in a transparent and robust manner, and – with an ABM – the modeler does not necessarily have to know every possible outcome in advance in order to model it well.

**Acknowledgements.** The author wishes to gratefully acknowledge his collaborators on the underlying research presented in this chapter. The marine ecosystem model presented in the first section was created in collaboration with Dr. Mirsad Hadžikadić; a fuller description of this work will appear in (Carmichael & Hadzikadic, 2013) later this year. Also, Dr. Charles Curtin provided important insights into the specific challenges of fisheries management and species reintroduction practices for ecological recovery. The second section on the energy use model of downtown Charlotte is one part of a larger project involving multiple models, as well as technologies for integrating these models into a holistic system of visual and analytical tools. This project is a collaboration between the Complex Systems Institute and the Visualization Center at UNC Charlotte, and IntePoint, LLC, a private software company co-located at UNC Charlotte. This project is ongoing, and the specific model presented here was developed under the management of Drs. Bill Ribarsky and Bill Tolone, and Mark Armstrong; and in collaboration with Kyle Thieman, Dr. Xiaoyu Wang, and Laura Hassey.

# References

Berger, J., Ridgeway, C.L., Fisek, M.H., Norman, R.Z.: The Legitimation and Delegitimation of Power and Prestige Orders. American Sociological Review 63, 379–405 (1998)

Bourdieu, P.: Distinction: A Social Critique of the Judgement of Taste. Harvard University Press, Cambridge (1984)

Brett, M.T., Goldman, C.R.: Consumer Versus Resource Control in Freshwater Pelagic Food Webs. Science 275(5298), 384–386 (1997)

Carley, K.: An Approach for Relating Social Structure to Cognitive Structure. Journal of Mathematical Sociology 12, 137–189 (1986)

Carmichael, T.: Complex Adaptive Systems and the Threshold Effect: Towards a General Tool for Studying Dynamic Phenomena across Diverse Domains. Ph.D. Dissertation. UNC Charlotte, Charlotte, NC (2010)

Carmichael, T., Hadzikadic, M.: Emergent Features in a General Food Web Simulation: Lotka-Volterra, Gause's Law, and the Paradox of Enrichment. Advances in Complex Systems (to appear, 2013)

Carmichael, T., Hadzikadic, M., Dreau, D., Whitmeyer, J.M.: Towards a General Tool for Studying Threshold Effects Across Diverse Domains. In: Ras, Z.W., Ribarsky, W. (eds.) Advances in Information & Intelligent Systems, pp. 41–62. Springer, Berlin (2010)

Chambers, D. J.: Varieties of Emergence (2002),
http://consc.net/papers/granada.html

Christensen, V.S., Guenette, J.J., Heymans, C.J., Walters, R., Watson, D.: Hundred year decline of North Atlantic predatory fishes. Fish and Fisheries 4, 1–24 (2003)

Coleman, J.S.: Foundations of Social Theory. Harvard University Press, Cambridge (1990)

Covey, R.: The 7 Habits of Highly Effective People. Free Press—A Division of Simon and Schuster, Inc., New York (1989, 2004)

Epstein, J.: Generative Social Science: Studies in Agent-Based Computational Modeling. Princeton University Press, Princeton (2007)

Festinger, L.: A theory of cognitive dissonance. Stanford University Press, Stanford (1957)

Flexner, J.T.: Washington: The Indispensable Man. Little, Brown and Company, Boston (1974)

Fromm, J.: Types and Forms of Emergence. Cornell University arXiv e-print service (2005), http://arxiv.org/ftp/nlin/papers/0506/0506028.pdf

Gardner, M.: Mathematical Games - The fantastic combinations of John Conway's new solitaire game 'life', vol. 223, pp. 120–123 (1970) ISBN 0-89454-001-7

Gladwell, M.: The Tipping Point: How Little Things Can Make a Big Difference. Little, Brown, Boston (2000)

Hadzikadic, M., Carmichael, T.: Overview of a New Approach to Economics. Swarmfest, Santa Fe, NM (2011)

Hannan, M.T., Freeman, J.: The Population Ecology of Organizations. American Journal of Sociology 82, 929–964 (1977)

Himoff, J., Skobelev, P., Wooldridge, M.: MAGENTA technology: multi-agent systems for industrial logistics. In: Proceedings of the Fourth International Joint Conference on Autonomous Agents and Multiagent Systems, Utrecht, The Netherlands, pp. 60–66 (2005)

Holland, J.H.: Emergence: From Chaos to Order. Perseus Publishing, Cambridge (1999)

Homans, G.C.: Bringing Men Back In. American Sociological Review 29, 809–818 (1964)

Jackson, J.B.C., et al.: Historical orverfishing and the recent collapse of coastal ecosystems. Science 293, 629–637 (2001)

Johnson, S.: Emergence: The Connected Lives of Ants, Brains, Cities, and Software. Simon & Schuster, New York (2002)

Kahneman, D., Tversky, A.: Prospect Theory: An Analysis of Decision under Risk. Econometrica 47, 263–291 (1979)

Kahneman, D., Tversky, A.: Choices, Values, and Frames. American Psychologist 39(4), 341–350 (1984)

Karlin, S., Taylor, H.M.: A Second Course in Stochastic Processes. Elsevier (1981)

Kennedy, J., Eberhart, R.C.: Swarm Intelligence. Morgan Kauffman Publishers, San Francisco (2001)

Khouja, M., Hadzikadic, M., Rajagopalan, H., Tsay, L.: Applications of Agent-Based Modeling to Pricing of Reproducible Information Goods. Decision Support Systems Journal 44(3), 725–739 (2008)

Lustick, I.S., Eidelson, R.J.: Secessionism in Multicultural States: Does Sharing Power Prevent or Encourage It? The American Political Science Review 98(2), 209–219 (2004)

Netlogo Web site, NetLogo is a multi-agent programmable modeling environment maintained at Northwestern University, http://ccl.northwestern.edu/netlogo/ (accessed on February 23, 2013)

Nisbett, R.E.: The Geography of Thought: How Asians and Westerners Think Differently—And Why. Free Press, New York (2003)

O'Brien, S.P.: Crisis Early Warning and Decision Support: Contemporary Approaches and Thoughts on Future Research. International Studies Review 12(1), 87–104 (2010)

Oksanen, L., Fretwell, S.D., Arruda, J., Niemelä, P.: Exploitation ecosystems in gradients of primary productivity. American Naturalist 118(2), 240–261 (1981)

Olzak, S., Olivier, J.J.: Racial Conflict and Protest in South Africa and the United States. European Sociological Review 14(3), 255–278 (2005)

Opp, K.-D., Gern, C.: Dissident Groups, Personal Networks, and Spontaneous Cooperation: The East German Revolution of 1989. American Sociological Review 58, 659–680 (1993)

Opp, K.-D., Roehl, W.: Repression, Micromobilization, and Political Protest. Social Forces 69, 521–548 (1990)

Ostrom, E.: Context and Collective Action: Common Goods Provision in Multiple Arenas. In: Héritier, A. (ed.) Common Goods: Reinventing European and International Governance, pp. 29–58. Rowman & Littlefield (2002)

Parsons, T.: The Social System. The Free Press, Glencoe (1951)

Poston, T., Stewart, I.: Catastrophe Theory and its Applications. Pitman, London (1978)

Power, M.E.: Top-down and bottom-up forces in food webs: do plants have primacy. Ecology 73(3), 733–746 (1992)

Rosenau, J.N.: Turbulence in World Politics: a Theory of Change and Continuity. Princeton University Press (1951)

Rosenzweig, M.: The Paradox of Enrichment. Science 171, 385–387 (1971)

Ross, C.E., Mirowsky, J., Pribesh, S.: Powerlessness and the Amplification of Threat: Neighborhood Disadvantage, Disorder, and Mistrust. American Sociological Review 66, 568–591 (2001)

Roy, S., Chattopadhyay, J.: The Stability of Ecosystems: A Brief Overview of the Paradox of Enrichment. Journal of Biosciences 32(2), 421–428 (2007)

Ryan, A.J.: Emergence is coupled to scope, not level. Complex 13(2), 67–77 (2007), http://dx.doi.org/10.1002/cplx.v13:2

Scheffer, M., De Boer, R.J.: Implications of Spatial Heterogeneity for the Paradox of Enrichment. Ecology 76, 2270–2277 (1995)

Taleb, N.N.: The Black Swan: The Impact of the Highly Improbable. Random House (2007)

Tarrow, S.: Power in Movement: Social Movements and Contentious Politics. Cambridge University Press, Cambridge (1998)

Tversky, A., Kahneman, D.: The Framing of Decisions and the Psychology of Choice. Science 211, 453–458 (1981)

Veblen, T.: The Theory of the Leisure Class. Viking, New York (1899, 1918)

Weaver, W.: Science and Complexity. American Scientist 36, 536 (1948)

Weber, M.: Economy and Society. University of California Press, Berkeley (1922, 1978)

Whitmeyer, J.M.: A Human Actor Model for Social Science. Journal for the Theory of Social Behaviour 28, 403–434 (1998)

Whitmeyer, J.M.: Prestige from the Production of Public Goods. Social Forces 85, 1765–1786 (2007)

Whitmeyer, J.M.: The Perils of Assuming a Single Equilibrium. Sociological Perspectives 52(4), 557–579 (2009)

Wilensky, U.: NetLogo Flocking model. Center for Connected Learning and Computer-Based Modeling. Northwestern University, Evanston (1998), http://ccl.northwestern.edu/netlogo/models/Flocking

# Author Index

Printed in the United States
By Bookmasters